Ultimate Guide to
Fishing South Florida
on Foot

7/9/2014

TO LEFTY KREH,

MY TRUSTED FRIEND,
WITHOUT WHOSE
UNWAVERING CONFIDENCE &
INVALUABLE ADVICE FEW OF
MY PROJECTS WOULD HAVE
COME TO FRUITION — THIS ONE!
ESPECIALLY.

THANK GOD TO
CONVINCED ME TO
COVER ALL
OF TACKLE TYPES

YOU

WITH
GREAT
BEST WISHES!!

Kent

Ultimate Guide to
Fishing South Florida
on Foot

STEVE KANTNER

HeadWater
Books

STACKPOLE
BOOKS

Published by
STACKPOLE BOOKS
5067 Ritter Road
Mechanicsburg, PA 17055
www.stackpolebooks.com

Printed in the United States of America

10 9 8 7 6 5 4 3 2 1

FIRST EDITION

Cover design by Caroline Stover
Cover photo by Adrian Gray with thanks also to Captain Ron Doerr

Library of Congress Cataloging-in-Publication Data

Kantner, Steve.
    Ultimate guide to fishing South Florida on foot / Steve Kantner. — First edition.
        pages   cm
    Includes index.
    ISBN 978-0-8117-1253-8
    1. Fishing—Florida.   I. Title.
SH483.K27   2014
639.209759'5—dc23

                                                                            2014009597

CONTENTS

PART IV: The Salt

ACKNOWLEDGMENTS

I wish to thank all my friends and associates who helped me make this book a reality. Lefty Kreh, who wrote the foreword, offered guidance throughout the project. Adrian Gray, of the International Game Fish Association (IGFA), lent his services by shooting excellent images and helping me arrange ones from other photographers. Many that appear here are from Pat Ford, who needs no introduction. Others were provided by Dr. Marty Arostegui, who holds more IGFA world records than any other angler in history (currently over 400), and Captain Mike Conner, of Stuart, Florida. Thanks also to Steve Waters, outdoors writer for the South Florida Sun-Sentinel, for sharing so much "insider" knowledge. His friendship and tenacity remain an inspiration.

An endeavor of this magnitude would have never been possible were it not for the input of everyday fishermen—ones whose triumphs often go unnoticed. However, standout performers like guide Alan Zaremba and tournament bass fisherman

Author slugs it out with a spinner shark from the privacy of a lonely beach. A variety of large gamefish can be caught from the surf in South Florida. STEVE KANTNER

Rene Gonzalez—both of whom helped me vet the freshwater sections—went out of their way to keep me on track.

The pier- and surf-fishing chapters bear the footprints of a legion of shore-bound anglers, but Rocky Fikki and Terry Luneke deserve special mention. My chapter on bridges would have been far less comprehensive had it not been for the efforts of Tom Greene and Captain Steve Anderson.

I'd be equally remiss not to thank the State of Florida, specifically several biologists who work for Florida's Fish and Wildlife Conservation Commission, including John Cimbaro (publisher of the *City Fisher*), as well as Paul Shafland (retired), Kelly Gestring, and Murray Stanford: the latter three of Florida's Non-Native Laboratory.

I've enjoyed an ongoing relationship with John Mazurkiewicz of Catalyst Marketing. John, who's also a fishing-rights activist, kept my efforts afloat with goodies from Shimano, Loomis, PowerPro, and previously 3M Scientific Anglers. For John's generosity I'm forever grateful, as I am to Ted Juracsik, who's helped me out for decades. Greg Block, of Jones and Company, who with his office staff, kept me stocked with X-Rap Rapalas and Chug Bugs—outstanding lures—along with digital images when I needed them. Mark Nichols of D.O.A. Lures did the same with his products, which I really appreciate. My thanks also to Jody Moore for the tidal data link and to Larry Dringus for the waterwatch link.

Thanks mostly to Vicki, my wife, who reread these pages until her eyes bled and who kept me on task when my resolve weakened. And, of course, to Andy Mill and Rebecca Wright, who took my calls or read a chapter whenever I needed their help. Those are the friends you never forget.

And last but not least, thanks to my editors at Stackpole—Jay Nichols and Tim Gahr in particular—for creating order from chaos while re-working this manuscript.

Steve Kantner
Fort Lauderdale
January 2013

FOREWORD

For decades, when anglers thought of South Florida, they have had visions of bonefish, permit, and tarpon in the Keys. But there is another fishing world that costs little to enjoy with spinning, plug, or fly tackle—and you can do it on your own by foot, car, or canoe. The number of species waiting for you is astonishing.

Steve Kantner of Fort Lauderdale gained fame as the Land Captain. With a canoe on top of his car, he cheerfully carried clients to such varied fishing holes as the interior of the Everglades, the Atlantic coast beaches and their piers, and the many canals meandering through towns nearly everywhere in the Sunshine State.

I have been lucky to fish with Steve in some of these places. The most unusual was in a canal coursing through downtown Fort Lauderdale where we caught

A new day is dawning for walk-in fishermen. This photo was taken near the Dania Beach Pier, along a stretch of shoreline where surf fishing is permitted when the lifeguards aren't on duty. Surf casters here target snook and pompano; pier fishermen, the same, along with Spanish mackerel, bluefish, and occasionally tarpon. PAT FORD

grass carp weighing nearly 20 pounds on a special fly Steve had developed just for this fishing.

Together we have paddled and fished canals for largemouth bass and bream and had a wonderful time. I especially enjoyed adventures when Steve drove us miles west from Miami on the Tamiami Trail and then took off on some forgotten dirt road penetrating into the heart of the desolate part of the Everglades. We'd fish abandoned canals for baby tarpon, snook, bass, and dozens of small species. It was pure fishing joy. We rarely saw another person.

After years of fishing on his own and then guiding professionally, Steve has investigated by modest means just about any water in which fish swim across South Florida, stretching from the tip of the south jetty at Fort Pierce to downtown Naples, the Glades, the surf, and even in the cities. He has at last assembled his huge store of experience and hard-won knowledge in *The Ultimate Guide to Fishing South Florida on Foot.*

TV meteorologist Jennifer Gray hefts a bonefish she landed on this shallow flat. Many flats are accessible to walk-in waders. PAT FORD

Steve is a master of all three types of tackle: spinning, plug, or fly. And he tells fascinating stories of his fishing adventures. But what is most useful are the detailed examples of how to rig tackle and what size hooks, lures, or flies to use, whether fishing the surf, the fishing piers, or paddling a rarely visited canal.

For anyone who wants to jump in his car with his kids or friends and drive to a place where he can fish on foot at very little cost, this book is the most useful piece of fishing gear you can own. It is invaluable for the hundreds of thousands of people who annually visit Florida and want a practical, down-to-earth guidebook detailing where, when, and how to fish countless nearby waters. It is the most complete fishing guide to South Florida ever published.

Good fishing!

—Lefty Kreh

INTRODUCTION

South Florida's diverse habitat provides shore-bound anglers endless possibilities as well as challenges. In this book I'll cover all the walk-in fishing on the mainland south of an imaginary line that stretches from the tip of the south jetty at Fort Pierce Inlet to a bistro chair in downtown Naples, not including the Florida Keys or saltwater venues that require a boat. That mainland consists of three separate zones: South Florida's interior, various predominantly saltwater venues and the brackish zone that I call the "in-between."

South Florida's interior includes nearly a million acres of designated Everglades—some of it located in conservation areas—plus portions of the Big Cypress and Corkscrew swamps. I list these wetlands together based on their proximity to each other as well as the species they support. This region consists entirely of freshwater habitat.

The in-between zone is a band of brackish and freshwater habitat approximately twenty miles wide that sits between the other two zones. This zone supports both native and exotic species, including migrants from both salinity extremes. Dikes and

The author, left, and Dr. Marty Arostegui hold up a brace of barramundi. Australian imports, barramundi were raised here briefly in a private lake for sale as food, as well as for sport-fishing opportunities—despite ecological concerns over their possible escape. Supposedly, none survived several recent cold fronts. PAT FORD

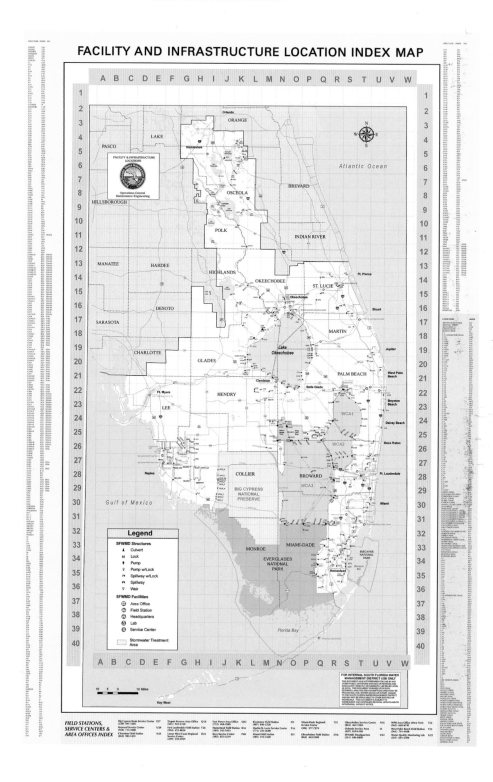

FACILITY AND INFRASTRUCTURE LOCATION INDEX MAP

Young Martini Arostegui, of Coral Gables, shows off a butterfly peacock he caught in a canal near his home. Peacocks can't tolerate any salinity. MARTY AROSTEGUI

levees separate it from the Everglades proper, while South Florida Water Management District spillways isolate it from the salt (see the maps starting on page 65).

As for saltwater venues, South Florida's beaches, inlets, jetties, ocean piers, bridges and adjoining seawalls, and any other bodies of salt water all lie on or near the Atlantic Ocean or, in some cases, Florida Bay. Fish from one venue frequently spill into others; it's a matter of salinity and similar habitats. To a lesser degree, it depends on the time of year. Each location has its idiosyncrasies, which is why each location requires different tactics and deserves a mention.

Delineations between regions take into account multiple factors, including salinity, tidal influence, and gravity-directed "sheet flow"—a product of the difference in elevation between Lake Okeechobee and Florida Bay. Take how real estate developers keep snapping up acreage. While a mixing of species is bound to occur, in wilderness areas especially, well-defined barriers prevent the introduction of, say, Mayan cichlids into Biscayne Bay or tarpon into Conservation Area Three. Or so we've been led to believe.

Some dividers—dikes and spillways, for instance—are strictly man-made, while others, like a given species' tolerance to salinity are natural. All these factors working together account for South Florida's dynamic ecological balance.

In order to fish successfully—especially on foot—you have to evaluate all these factors, while figuring in the weather and seasonal changes. Although a few well-known honey holes have consistently good fishing, the majority change with the

A lone spin fisherman taps the surf with an ocean pier in the background. Surf casting offers anglers solitude, along with the possibility of excellent fishing. PAT FORD

tide, the amount of recent rainfall, wave action, and wind direction—anything that could influence a fish's behavior.

While some anglers yearn for hard-and-fast answers (the kind they get from a GPS), you won't find them here, at least not in the traditional sense. Instead, I'll teach you to interpret the dynamics so you stay one step ahead of the charts and the website bulletin boards.

According to state biologist John Cimbaro, Florida has three million acres of lakes, ponds, and reservoirs, plus 12,000 miles of rivers, streams, and canals. Many of these are located in the state's southeastern corner. The beauty of the fishing opportunities described in this book is that for the most part you don't need a boat to access these waters (although a small craft such as a canoe or kayak may be helpful in certain situations). That amounts to a substantial savings, especially for traveling anglers.

Many of our inland waters parallel roads or cut through shopping malls, while others literally abut our homes. While some run fresh, others turn brackish—or salty—according to the dictates of rainfall and tide. What inhabits them varies according to the salinity, but anglers who fish them are frequently surprised to find both fresh- and saltwater species obliging on successive casts.

There's more to walk-in fishing than just narrow waterways. Take all our accessible inlets and saltwater seawalls. Then there's the River of Grass, which offers world-class bass fishing.

Most expansive of all is the surf. Florida's beaches are proof that not all good fishing is confined to wilderness settings, and not all urban venues have been swallowed by crowds. Even after years of remorseless development, a mind-boggling array of hot spots still exists within metropolitan areas as well as along major high-ways. In fact, some of the fishiest roads lead to skiff destinations.

Anglers along the coast fish from shore, jetty, and pier—equipped with every-thing from spinning gear to gargantuan winches. These anglers battle an assortment of run fish as well as giant snook and jacks, sharks of all sizes, monster tarpon, and oceangoing pelagics, including everything from false albacore and smoker king mackerel to the occasional tuna, dorado, or sailfish.

In the following pages, I share what I've learned on the water over more than fifty years. In all the years I fished as the Land Captain, I guided clients to venues that ran the gamut from swamps to surf. I'll cover a wide range of fishing tackle, from fly fishing to conventional gear, and share advanced tips for experienced anglers as well as fundamental advice for newcomers.

Because I don't mind admitting when I'm unsure of a detail, I've sought the advice of other anglers. Some names you'll recognize; others you won't because certain individuals—although experts in their fields—eschew the limelight and fish for fun, which after all is the point of this, isn't it?

Steve Kantner
Fort Lauderdale
January 2013

It's back to the parking lot at Blowing Rocks Preserve, with me on the right. The surf between Fort Pierce and Juno Beach hosts seasonal runs of bluefish, pompano, Spanish mackerel, and herds of small jacks, along with larger predators, such as snook, oversize crevalles, and spinner sharks. Hutchinson Island surf casters recently enjoyed a mid-summer run of false albacore.

MIKE CONNER

Fishing South Florida

PART I

Current, Seasons, and Weather

Current is essential to most types of fishing; it brings prey to predators, and it seems to calm the fish. The current typically comes from several sources: tides, water releases, and sheet flow. Tides are mostly the result of celestial forces, while spillway releases come from a buildup of runoff. Sheet flow is a product of gradient (hence the term "River of Grass").

The bridges of the Overseas Highway, where techniques such as dropping (and recovering) rods and splicing lines quickly helped my friends and I land numerous huge tarpon. Here, as on the mainland, the tides set the pace, as predators wait near the pilings for current-borne forage.
PAT FORD

The Palm Aire Spillway (S-37B) in Pompano Beach—historically, a source of midsize tarpon. Current is critical to good fishing. Immediately after a thunderstorm, certain canals beneath spillways become stuffed to the gills with tarpon and snook feeding on freshwater forage fish that get washed through the gates. STEVE KANTNER

Tides

Tides are all-important when you're fishing the salt. The moon and the sun exert gravitational forces on large bodies of water, which account for the rise and fall of the tides. Tides here in South Florida, from Palm Beach to Naples and south through the Keys, fall and rise (we say ebb and flood) according to a semidiurnal cycle. That means two high and two low tides during each tidal day. A tidal day is 24 hours and 50 minutes, which is how long the moon takes to orbit the earth.

While some tides are higher or lower than others due to the phases of the moon, fluctuations in this region typically average between one and three feet. Stronger spring tides, which occur during and immediately following full and new moon periods, have nothing to do with the time of year. Instead, they result from increased gravitational forces that occur when the moon and sun are in direct alignment, which occurs twice a month.

Predators and Current

Predators enter the shallows at the start of incoming tide and feed "uphill" as the water rises. They'll hang out in skinny water at the top of the tide before backing off when it starts to drop. Some predators wait out the drop in bays or channels, while others stake out creek mouths or points of land, where they ambush forage fish that are swept downcurrent. Snook are a prime example of a species that feeds mainly during falling water.

Something similar takes place in the surf, albeit for opposite reasons, when the deeper water afforded by incoming tide encourages predators to feed closer to shore. All fish, be they minnows or 100-pound tarpon, fear attacks from above from seabirds or land-based predators. Their sense of security—heightened at dawn and dusk—ramps up further in the presence of current.

Fishermen line the Jupiter Jetty in order to take advantage of the incoming tide. Fishermen on jetties rely on current, especially the falling tide. That's when everything darkens, like it does in the salt marsh, which encourages predators to forego their caution. Added camouflage fuels the predators' ardor—more so at night, but during daylight, too. STEVE KANTNER

We also experience weaker neap tides around the moon's waxing and waning quarters. They result when the sun and moon are at a 90-degree angle, with the earth at the apex. Tides in southwest Florida—for the sake of this book, the salt marsh—are often referred to as "mixed." This means that one high and one low per tidal day are higher or lower than their most recent counterparts.

A strong offshore or onshore breeze can affect the tide. Barometric pressure also influences tides, since the weight of a heavy air mass (say, more than 30 inches of mercury like Florida experiences during winter cold snaps) pressing on a large body of water suppresses both high and low tides.

Low pressure (like during tropical cyclones) allows the water to rise, so both the high and low tide are appreciably higher. The more water that moves, the stronger the current, which improves the fishing. These tidal fluctuations can be felt several miles inland, although they won't be as noticeable near the coast. Nonetheless, they affect the fishing.

Sheet Flow

There's even current in the Everglades. Glades historian Marjorie Stoneman Douglas called the 25-mile-wide swath of saw grass and cattails that stretches from the southern border of Lake Okeechobee (give or take a mile or two) to the northern-most limits of Florida Bay a "River of Grass." South Florida isn't flat; in fact, a

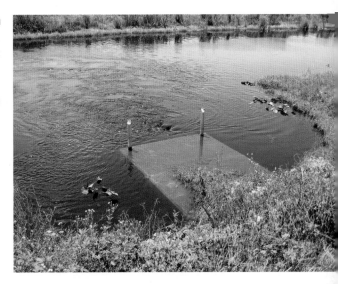

A tiny culvert directs water through an Everglades levee. It's all part of sheet flow, which reaches its peak in the spring. Predators, large-mouth bass especially, lurk just under or outside the current, while waiting for forage. MARTY AROSTEGUI

12-foot gradient separates Lake Okeechobee from Florida Bay, which accounts for sheet flow, or surface water trickling in a southerly direction. The Glades drain southward toward Florida Bay. While the highest elevation, in Palm Beach and Broward Counties measures in the low double-digits in feet, that's enough to keep the water moving.

The Everglades is just a flooded pasture that's intersected by man-made canals. While you won't find much current on the actual prairies, it's frequently visible beneath the region's bridges or in north-south canals. And like rivers everywhere, it begins in the sky.

Seasons

Florida has two main seasons—rainy and dry. The first, and most dramatic, is the rainy season (or hydroperiod), which roughly corresponds to the Atlantic hurricane season. It's during this five-month spate of heavy rain activity, typically beginning in mid-May, that nearly three-quarters of the region's 60 inches of average annual precipitation falls. Those afternoon cloudbursts are violent events, with hourly rain-fall rates that reach several inches. I watched 11 inches fall late one afternoon, iron-ically after leaving South Florida Water Management Headquarters. Tropical cyclones can drop a foot or more of water in a matter of days. Locals call it the summer pattern, and it's when temperate breezes are considered the norm. A sec-ond, drier season (the drought) extends from the first fall cold front until sometime in May. Only one-third of the region's annual precipitation falls during the drought.

The Rainy Season

On an average June day, the prevailing southeasterlies pick up moisture as they cross the Gulf Stream (or drift over the Everglades). Then, driven by afternoon heat,

While prevailing breezes here blow from the east, the weather systems themselves follow the opposite track while riding the jet stream, a high-speed river of air that circles the Northern Hemisphere from west to east. In June the rules reverse temporarily—at least, as far as tropical cyclones. Hurricanes generally form in the tropics or Gulf, although this is not always the case.
PAT FORD

the moisture rises and starts to condense along what locals refer to as the "sea breeze line." It's an ageless process that refuels our aquifer.

Most thunderstorms spawn over the state's interior. Take the following example: It's 2:00 pm on a balmy day, and those puffy clouds of a few hours earlier are mushrooming into pillars of lethal convection. Inflows and outflows (visible appendages of the developing storm) transform the sky into a palette of grays—and occasionally greens if a tornado's forming. Then a few minutes later the fireworks cut loose as the storm takes aim on the coast. Thunder crashes and lightning flashes before the ensuing downpour brings life to a standstill. Once it's over, quiet returns to a backdrop of ankle-deep puddles. Their energy spent, the storms drift off to sea—where eventually they'll vanish into oblivion—while the fish get back to the business at hand.

The redistribution of all that moisture triggers spillway releases as well as lesser flows. One way or another, rain creates current in which gamefish gather to feed, and the cooling effect invigorates fish and man.

These storms have a sinister legacy of the deaths caused by lightning. Runoff from the rains also hurts our coastal fisheries. Anyone planning to visit during the rainy season should avoid these storms whenever possible. Fish early or late, and don't take chances.

The fireworks let up by November, when daytime temperatures become pleasantly tolerable. Whenever they dip into the mid-70s, the downpours come to a halt, for the most part—even though we can have dangerous squalls in the winter.

The Dry Season

When the drought season begins in earnest, usually by February, the swamps really start draining. Together with the drop in daytime temperatures, the receding water triggers explosive fishing. As the days roll by, the dry spell worsens, until May when it hits its peak. By then, all vegetation (including lawns) starts turning brown, while watermarks appear at the base of the cattails. Some sunken prairies have been reduced to puddles. Find one on the verge, and you'll see wading birds by the hundreds or thousands gathering to feed on all the stranded baitfish—most of them mosquitofish (also known as *Gambusia*).

The drought forces forage to forsake the marshes and head for the safety of deeper water where it's vulnerable to attack from a new list of predators. Some baitfish and grass shrimp don't make the transition and succumb to the drought or the birds. Still, enough find their way into creeks and canals to provide grist for the gamefish that anxiously await them—predators that sensed their arrival on the quickening current. You can view the process throughout the interior (as well as in parts of the in-between), since most inland waters have similar fluctuations. In the salt marsh especially, new species keep arriving as the glut of forage sweeps into creeks and canals.

A prairie drains as winter approaches.
STEVE KANTNER

Wading birds gather where forage enters a creek or canal. STEVE KANTNER

This trophy largemouth hit a sinking fly while the water was dropping. Predators take advantage of current, especially while water levels continue to fall. MARTY AROSTEGUI

As the water keeps dropping and predators start competing, the blitz is on. One of the main clues that the blitz has started is that you see showering rain bait: tiny baitfish that hug the shoreline. Look for jumping bait along sunny shorelines, near floating docks, and in the vicinity of launch ramps. It won't be long until you see swirls erupting, typically (but not always) near emergent vegetation. Then, before you know it, the bite's in full swing.

April and May are Florida's driest months, when land that was recently submerged sits high and dry. Last year's runoff has done one of three things: stoked the aquifer, mixed with the salt, or refueled the clouds through evaporation. When the rains begin, the cycle renews, and the drought is once again a thing of the past. For now, however, the fish couldn't be happier.

Cold Weather

Cold in South Florida can do irreparable damage, in salt water as well as fresh. It pushes migratory fish south in large numbers, while more sedentary species (such as snappers and triggerfish) float to the surface when trapped by falling temperatures. The cold hits some inshore species the hardest. Snook, for example, may be stunned or killed if trapped in the cold for more than a few hours. Some head offshore, some swim upriver to a freshwater spring, and some gather in deepwater inlets or below power plant discharges.

While stay-at-home species are the usual casualties, powerful swimmers may also fall victim when the chill persists for days at a time or in shallow areas that are low in oxygen. Tarpon and sharks are two examples. Meanwhile, redfish, seatrout, black drum, and sheepshead—all of which tolerate cold—fare much better.

As far as the salt marsh, whenever nighttime temperatures in the backcountry dip below 50 degrees, fishing is iffy at best. I plan my fishing based on National Oceanic and Atmospheric Administration (NOAA) forecasts, which are right on the money. Look for the projected nighttime lows in Naples before making your decision. Or avoid the hassle entirely by relegating questionable days to the surf or a pier, where cold-hardy species may be anxiously waiting.

Coldwater Eddies

Coldwater eddies have little to do with cold weather—in fact, just the opposite. Whorls of cold water spin off from beneath the Gulf Stream, where a cold counter-current flows in the opposite direction. These plumes of cold water work their way inshore, where they ruin inshore and reef fishing. Nowhere is this more apparent than in Palm Beach and Martin Counties, especially during the summer.

Immediate effects include the appearance of deepwater species, such as vermillion snappers, in shallow water, and an unscheduled run of undersized cobia—typically males—along Treasure Coast beaches. There's always the chance of a legal fish. As usual, most cobia will be following stingrays.

During coldwater upwellings that affect Hobe Sound Beach, snook, tarpon, and other tropical species vacate the surf. Yet the snook may remain behind, especially if temperatures in the Indian River Lagoon are unseasonably low for that time of year. On the other hand, when temperatures in the St. Lucie and Indian Rivers drop too low, the snook in the rivers (easily identifiable by the yellow color in their body and fins) head for the beach regardless of surf temperatures.

Essential Flies and Lures

An inviting aspect of most local fisheries is not their diversity, but the parity that exists between, say, mangrove creek and manicured seawall, or inlet and pier. Despite differences in location, surroundings, or even salinity, prey found in varied environments looks surprisingly similar to the human eye and, apparently, to the fish as well—if you exclude garish contradictions like ficus berries and acorns. So lures that are effective in one location are often effective in others, too, offering simplicity and savings for savvy anglers.

Fly patterns for the in-between, starting from left: Modified Clouser Deep Minnow, purple Flashabugger, modified Squid Fly, Rivet, black Marabou Muddler, pink Suescun's Conehead, pencil popper. PAT FORD

Both the yellowbelly and the Mayan cichlid figure in the local food chain. Brightly colored Clouser Minnows imitate these and other cichlids. KELLY GESTRING/FLORIDA FISH AND WILDLIFE CONSERVATION COMMISSION NON-NATIVE LAB

Gambusia, or mosquitofish: universal forage from suburban canals all the way to the brackish hinterlands of southwest Florida. Imitate these tadpole-shaped baitfish with Mike Conner's Glades Minnow, Lou Tabory's Squid Fly (modified), or the Rivet: All the best imitations have an expanded head. KELLY GESTRING/FLORIDA FISH AND WILDLIFE CONSERVATION COMMISSION NON-NATIVE LAB

Inland Waters

Tiny baitfish that measure less than two inches long are the mainstay here for largemouth, peacocks, tarpon, and snook. While the scientific names of these baitfish vary, most resemble the potbellied guppies and mollies found in children's aquariums—think rounded bodies and flattened heads and fish that swim on top like a tadpole. That forms a solid basis for imitation.

Threadfins and their larger cousins gizzard shad are swept downstream by spillway releases—triggering unbelievable fishing for tarpon and snook. They're also important in certain lakes—like Ida in Delray Beach—where they provide predictable forage for predators. Wobbling crankbaits, with and without lips, do a credible job imitating these prey.

Golden shiners are another option in suburban waterways. They're the go-to bait a bit farther upstate, where trophy hunters pursue outsized largemouths.

While shiners are available at several freshwater bait shops, you can catch them yourself by chumming with oatmeal before throwing a cast net or by suspending a tiny bread ball beneath a bobber. The same lures that imitate shad imitate shiners.

While juvenile sunfish and cichlids, with their rounded shapes and flattened silhouettes, have a place in the subaqueous food chain, only a few predators devote time to pursuing them. Butterfly peacocks and snook occasionally target tiny cichlids, while largemouth bass prefer baby bluegills. When you encounter this kind of selective feeding, you'll want an accurate imitation like a slab-sided lure—especially when dealing with backwater snook. While the types of imitations are almost limitless, you can downsize your tackle if you follow a few rules: Limit the number of lures you carry, while keeping in mind guppies and mollies. A few unweighted shads offer one-stop shopping, as long as they're not too large. As with all soft plastics, fish hang on longer. Marabou crappie jigs are another option, whether you fish the Glades, the salt marsh, or off a pier or a bridge; however, their light wire hooks have a tendency to straighten on a good-size fish. This is especially noticeable in the salt marsh, where 20-pound snook ingest inch-long forage. Very important when selecting lures is that salt marsh gamefish key on subaqueous signals, which is why flies and lures that push water work so well.

March of the Frogs

Even amphibians occasionally migrate, which reminds me of a story told by Captain Bill Curtis. It's set back in the days when the late Rocky Weinstein was heralded as the original land guide. The way Bill tells it, Weinstein waited for his charters to end, and then, if he'd seen a trail of squashed frogs on the highway, he'd call Curtis and Chico Fernandez.

The pair would rush over from Miami and wait behind the cattails along Route 29 for the tide to start dropping. The frogs had established routes—hence the rows of carcasses. The snook, in turn, had figured this out and took up positions where they laid in ambush. Once the carnage began, the snook would hit anything that even remotely resembled a frog. Curtis showed me a snapshot of a 27-pound snook that he landed on a plain white slider.

I've enjoyed spectacular fishing with plastic frogs—especially for largemouth bass with bait-casting gear. STEVE KANTNER

If the predatory antics of snook, bass, and tarpon can be called "explosive," then the softer approach of tilapia, mullet, and Chinese grass carp—all vegetarians—is truly bizarre. Imagine casting to fish that are slurping salad: take the fruits that drop from bankside ficus trees or floating blossoms. Tapping into this fishery is new ground. Since imitating vegetation is still in its infancy, anglers and fly tiers have little to go on. If the prospect sounds daunting, visualize berry imitations and synthetic bread. PAT FORD

In addition to baitfish, insects, crustaceans, and even berries can be important. While I've never seen bass gulp caterpillars, ants, or beetles, on many occasions I've watched them clear the surface in pursuit of newly fledged dragonfly and damselfly adults. Fly fishers benefit when these bugs are active by fishing poppers or Woolly Buggers near emergent vegetation.

You'll find shrimp-like scuds (Amphipoda) in floating vegetation as well as clouds of grass shrimp during low-water periods. Crustaceans can incite a feeding frenzy: Just visit the swamp on a quiet evening and listen to the bluegills slurping scuds. You'll learn to recognize when fish are eating grass shrimp (or scuds); the popping sound is a major tip-off. Meanwhile, crustacean feeding cries out for the accuracy of a specialized fly.

In salt marsh country you hear the same sounds, but with the volume turned up so that it sounds like a boom box. Snook will be banging away beneath the bridges—another example of selective feeding.

Topwater Lures for Fresh Water

Topwater lures are my hands-down favorite, and most other anglers agree—whether they're spin, plug, or fly fishermen. A topwater lure elicits savage strikes but your success rate is subject to your quarry's whimsy, meaning if conditions aren't right you can't buy a strike. Other species besides largemouths hit surface baits, too, including a plethora of panfish as well the ubiquitous gars and mudfish.

When fishing surface lures it usually pays to concentrate on shorelines—drop-offs, too, if you know where they are—usually between 3 and 10 feet from shore, depending on the water depth and slope of the bottom. Emergent vegetation also harbors fish, as does the stuff that floats—swamp cabbage, in particular. Rocky shorelines where forage can hide rank among the best places to fish—forget what you've heard about a lack of vegetation. That's because some Everglades bass are constantly moving, patrolling the banks. So when a barren stretch suddenly yields three or four bass in a row, it may have nothing to do with changing lures, but sometimes it does. When you start thinking of bass as far-ranging predators, you'll improve your totals.

The typical retrieve for surface bait incorporates a reel-and-stop cadence (or strip-and-pause with a bass bug), interspersed with pauses of varying length. The trick lies in the timing. I've tried letting a popper lie still while waiting for the ripples to die. While it's the optimal way to target bluegills, a more spirited retrieve outperforms it for bass—with two notable exceptions.

The first exception comes after a period of rising water, usually following a spate of cooling rains. Whenever that happens, the bass go deep and suspend along drop-offs. The second exception? Bright overhead sunlight late in the season, when the bass start scurrying for deeper water—say, along that same drop-off. The first situation may catch you off guard, since the bass were hitting just a few days earlier. I attempt to solve it by casting a loud plug or popper to wherever I think there's a drop-off and allowing it to sit still for at least 10 seconds before giving it a whack by twitching my rod tip.

What makes one presentation more effective than another? It probably has to do with triggering impulses, with the nod usually (but not always) going to what's new and different. While hunger is definitely a trigger, territoriality is also important—as is curiosity, along with just plain cussedness. But it comes down to more than that, considering how bass are so willing to change their preferences.

Still, certain rules nearly always apply. Bright finishes, such as fire tiger (a finish marketed by numerous manufacturers) and chartreuse, excel in murky water, while more subtle shades out-fish them in clear. Fire tiger, a medley of orange, black, and green, mimics the baby cichlids that infest our waterways. So you end up with something that looks like forage but also resembles an Andy Warhol painting. When you're fishing an environment that's devoid of cichlids and where the water is noticeably clear, stick with the drab stuff: gold or silver sides and a black or brown back—or possibly an all-clear finish. It's not just what gamefish see, but what they don't see that can make a difference. Realism, unsurprisingly, is always in style—which explains why frog finishes are perennial favorites.

But it's how we work these artful creations that trips the fish's trigger. Surface lures, although similar in appearance, have idiosyncrasies that set them apart. It may have to do with size or color, but it's more likely the result of the signature splash they make when you pop them during retrieves or how much water they disturb.

Technique is critical to success, and certain tricks can increase your chances. But the key for surface lures is to know the time and place to use them. Focus on fishing at dusk or dawn, or during other low light periods when bass are eyeing the surface.

The following list is my favorite surface lures and techniques for using them. Among my all-time favorites are Storm's Chug Bug, Rebel's Pop-R, Heddon's Zara Spook, and Smithwick's Devil's Horse, as well as several sizes of floating Rapala—all of which float at rest.

The ***Chug Bug Popper***, a cupped-face floater manufactured by Storm, makes a commotion when twitched on the surface. But it's still small enough for yearling bass, and it's what I start out with if I'm planning to spin-fish. I retrieve it like I would a fly rod popper, with jerks and pauses in syncopation, or experiment by speeding up or slowing down my retrieve.

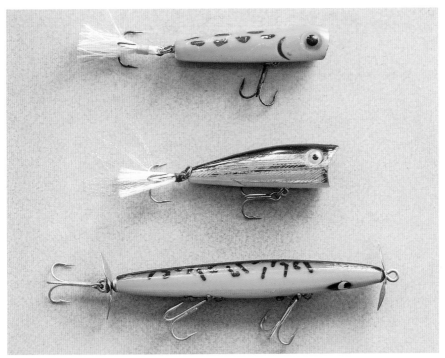

Popular surface baits for largemouths, from top to bottom: Storm's Chug Bug, Rebel's Pop-R, and Smithwick's Devil's Horse. STEVE KANTNER

The Heddon Zara Spook, a leading topwater bait for more than half a century, is still turning heads in freshwater lakes and canals.
STEVE KANTNER

The **Pop-R**, a fat, cup-faced floater from Rebel, is even louder. I save this plug for when I want to make an unmistakable noise, like when bass go deep after a cool spring rain, or later in the season when the surface sizzles. I cast the Pop-R beyond where I think there's a drop-off and give it a whack with my rod. Then I'll pop it again when the ripples die, after sitting on my hands to pass the time. I wait thirty seconds for that second whack and repeat the process another time or two. I'll keep this up for several minutes before retrieving and casting again. The strikes, often subtle, usually come out of nowhere.

Heddon, the company that introduced the classic hard bait, the **Zara Spook**, later followed up with the Baby Zara, the Zara Puppy, and the Zara Pooch—all smaller versions of the same design. At rest, the Zara Spook sits head up on the surface; twitch it and it immediately bobbles. The larger sizes, which I find more effective for

Floating Rapala in natural and Fire-tiger finishes. GREG BLOCK/JONES AND COMPANY

bass, are at their best when I'm "walking the dog"—that swaggering, side-to-side motion I get by moving my rod tip back and forth while reeling and pausing. Holding my rod tip 45 degrees off the water makes this easier. This is standard procedure for working a surface bait.

When it comes to hard baits, you can't beat a ***Rapala***—particularly the floaters—which in sizes 3 through 7 are possibly the world's most popular twitch bait. You can also fish Rapalas crankbait style by picking up your retrieve and keeping it steady. For several weeks each spring before the water warms, this plug is the most effective bait in the entire Everglades.

Try tossing one near some floating swamp cabbage and allow it to sit. Then make it bobble by twitching your rod tip. In the clear, tannic water of the Glades, natural shades—including nondescript finishes like rainbow trout—get the nod. If the water's off-color, then the brighter shades like fire tiger or chartreuse work better.

Subsurface Baits

Not all great bass baits make a ruckus on top. In fact, the lure that changed the face of bass fishing—Nick Creme's plastic worm—was originally retrieved near the bottom. Meanwhile, subsurface bass baits come in two basic styles: soft and hard.

Soft Plastic Bass Baits

While most soft plastics can be reeled on top (or just beneath the surface), others—like the Senko—were designed to sink. Slow-sinking baits run the gamut

Plastic worms of various sizes, shapes, and sink rates have revolutionized bass fishing. STEVE KANTNER

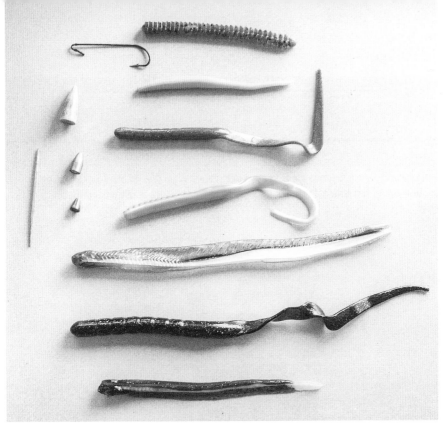

Plastic worms, a classic example of soft baits, are available in numerous shapes and styles—ripple tail, curly tail, paddle tail, or French fry—and all can be effective at times. Trick or finesse worms are extremely thin worms that respond to the slightest twitch, which makes them ideal for light lines and spinning gear. Various worms, from top: French fry, tapered trick worm, ribbon-tail worm, curly-tail worm, slug, magnum ribbon-tailed worm, mini-slug. STEVE KANTNER

Texas-rigged worms, like the one in this photo, account for a major percentage of bass caught on soft baits. PAT FORD

Wacky Worm fishing involves hooking the bait in the clitellum (the fat band just ahead of the mid-point), and letting it relax on the retrieve between twitches. It's a deadly technique.
STEVE KANTNER

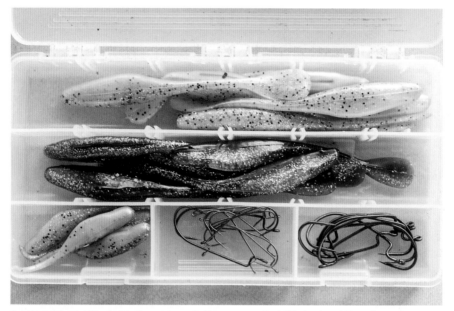

Gambler Flappin' Shads (the larger lures in the upper and middle tray) and Bass Assassin Original Shads are deadly in fresh water. In my estimation, nothing out-fishes a Bass Assassin when bass are feeding on baitfish. (My favorite is the original 3½-inch shad, rigged weedless.) You can fish soft baits bare or behind a jighead. If you choose the latter, use either a straight retrieve, or bounce them slowly along the bottom, depending on where the bass are holding.
STEVE KANTNER

from worms, crawfish imitations, and so-called shads to creature baits that imitate frogs or lizards, and all types of forage in between.

You won't need much weight with these soft baits—except to break through the crust or if you intend to "flip." But if you fish a soft plastic in conjunction with a worm lead (and you aren't using a screw-in weight), pin the lead in place with a toothpick. Insert it in the tip of the lead before breaking it off. Adding a drop of superglue also helps.

A jerk bait is similar to a twitch bait, but you jerk it harder than you would a twitch bait. I'm partial to plastic shads. These baits, which were intended to resem-

Rapala Shad Rap is a highly effective suspending crankbait. Tip: If your crankbait pulls to one side, bend the eyelet gently in the opposite direction. STEVE KANTNER

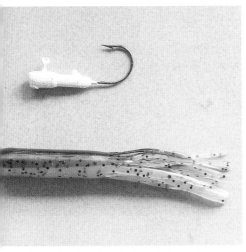

Tube jig with elongated head. These lures excel not only for panfish, but also for bedding largemouths. STEVE KANTNER

Modern bass hooks come sharp from the factory, with offset models being among the most popular. As a general rule with plastics, bury the point just beneath the surface so it rips out easily when you set the hook. STEVE KANTER

ble their namesake, are typically retrieved with a series of jerks and pauses. A product known as the Sluggo was the prototype but, my current favorites are the 3½-inch Bass Assassin and the larger, but similar Flappin' Shad.

The Flappin' Shad, which is manufactured locally by Gambler, is the perfect example of an all-around lure. Reel the 6-inch version across the surface and it's a full-size attractor. Bite off its nose and it becomes a popper, or tear off its tail and it's an effective twitch (or jerk) bait. Add enough weight and you're ready to flip. What, exactly, is flipping? Well, if you add enough weight, you can break through anything, including dense vegetation, and catch bass that are lurking beneath it.

You'll want heavy line, though—say 30- to 50-pound braid—to derrick your catch through the sunken salad.

Crankbaits

Crankbaits, as their name implies, need only to be reeled to be effective. Some have lips (like the Fat-Free Shad or floating Rapala), while others do not (like the popular Rat-L-Trap). Back a few years ago, I started seeing fatter designs that looked more like a teardrop than a shad or bluegill. Still, all wiggle seductively like an injured baitfish during retrieves.

Jigs

Jigs, a traditional subsurface bait, are effective in a number of situations. They're deadly below flowing spillways, where nothing lighter could get through the garfish. The experts add a strip of pork rind or a portion of rubber worm to enhance their jig's action (called "a jig and a pig").

Tube jigs, which vaguely resemble a saltwater squid, are highly effective at certain times, like when largemouths and crappies are spawning. Why would a bass or crappie grab such a strange-looking lure? It must have to do with its daunting appearance. Insert the elongated jighead into the hollow body before poking the hook eye through the plastic skin.

Freshwater Flies

The topwater and subsurface patterns that follow will cover most situations. Patterns from my salt marsh box—the Rivet, for example—are effective if the bass go on a killifish rampage. Certain in-between patterns such as Mike Conner's Glades Minnow, Bob Clouser's Deep Minnow, and the Marabou Muddler also work well in the state's interior. Additional keepers include a ficus berry imitation for grass carp.

Cork-Bodied Popper

The cupped-face, cork-bodied popping bug is the unofficial fly of South Florida's interior. Although originally designed to lure smallmouth bass, cork poppers are also effective for largemouths, along with bluegills and other native panfish, and several exotics.

A well-made bass bug is a thing of beauty, and cork is the material of choice for tiers like me, who cling to the past while striving for utility. Various closed-cell foams run a close second, but they fail to provide the same buoyancy as natural materials.

The most popular colors in the Everglades and surrounding areas are pearl, black, gray, robin's egg blue, carrot, and chartreuse. Most bass fishermen attempt to "match the hatch"—or at least choose colors that show up more easily depending on factors like glare and water clarity, and they change colors based on the fish's preferences.

Let's go fly fishing: Fly box full of freshwater patterns, including assorted poppers, streamers and a fly rod "lure" (The Finished Bug) that's often effective when the others fail. STEVE KANTNER

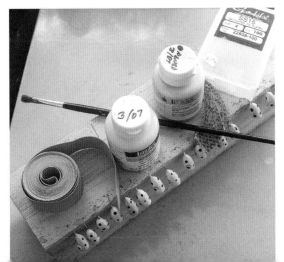

I examine the corks for defects before gluing each hook in the slot with five-minute epoxy. Once the epoxy sets, I coat each body first with Gesso (available at an art supply shop) and later with colored acrylics. I use pin heads to apply both the eyes and the spots. Then, when everything dries, I apply acrylic gloss.

STEVE KANTNER

I prefer weedguards made from a hard mono loop, or horseshoe, that extends from the lure's underside. If you purchase a bug without a weedguard, you can sew in a loop of Mason's 10-pound hard mono before gluing it in place with a dab of superglue. While weedguards are essential for fishing around cover, they shouldn't interfere with a bug's hooking ability, so be sure that they clear the hook point when pushed against the body.

I seldom go smaller than a No. 6, or even No. 4 as the season progresses. Then, when summer arrives, I'll switch to a Legs Diamond hairbug.

Cork-Bodied Popper

Cork-bodied poppers in various colors are an Everglades staple. While chartreuse is arguably the most popular color, gray and powder-blue are also in the running. Blue, incidentally, imitates a particular dragonfly.
STEVE KANTNER

Hook:	Gamakatsu SS15
Thread:	Flymaster Plus
Body:	Cup-faced cork, Gesso, colored acrylic, clear acrylic, and five-minute epoxy
Tail:	Calf tail, stacked, 1½ times body length
Flash:	Pearlescent Krystal Flash
Hackle:	Webby, poor quality grizzly saddle, dyed to match body.
Legs:	Rubber Legs
Weedguard:	10-pound Hard Mono horseshoe

Notes: Make a few turns of hackle to cover the spot where the tail was added. Tie in three or four legs before you add the tail; then, sew two more through the body after completing the other tying steps.

Flashy Buggers

In the Everglades, where the varied forage—from dragonfly nymphs to *Gambusia* baitfish—may not be visible in industrial-size quantities, matching the hatch is a crapshoot. That's why I look for patterns that cover multiple bases, and no pattern imitates more things than a Flashabugger.

I have no idea who tied the first of these flies, but they're a souped up version of the Woolly Bugger streamer, which, in turn, evolved from the Woolly Worm wet fly. Woolly Worms are typically tied with their hackles facing forward—plus, they either have no tail or the tail is a short yarn or hackle stub—while the Bugger's hackles face to the rear. Buggers are a lot more streamlined and behave at their best when stripped deliberately. All have marabou tails that are at least body length.

Woolly Worms were once popular on Lake Okeechobee, during spring when monster bream spawned in the shallows. I fished with several guides out of Moore Haven, Florida, who could locate the beds by smell—no kidding. After dropping my fly directly on top of a bed, I'd let the bluegills and shellcrackers (redear sunfish) do the rest. I once landed a mixed stringer of 48 that weighed just over 50 pounds.

I know of no fish that won't hit a properly placed Bugger. As a case in point, I've enjoyed unexpected fishing for offbeat species, such as plant-eating tilapia. While our native panfish assault this pattern, oscars and Mayan cichlids can also be downright brutal. Freshwater snook hit Flashabuggers, too—not surprisingly, the darker versions, which no doubt remind them of the ubiquitous black mollies that infest roadside canals.

More than anything else, this pattern offers flexibility. Since it represents a style of tying, as well as a pattern, choose whatever color combo you wish, while keeping in mind that the darker shades—which imitate insect larvae or tiny minnows—are more effective locally than some more colorful versions. Purple—which excels for both tarpon and peacocks—still ranks among my favorite colors—but you can't beat a black one in the state's interior.

Black remains my go-to color for these quiet backwaters. It's also a contender when I fish the salt marsh, and there's no visible feeding. Allow it to sink near emergent vegetation before beginning a slow-twitch retrieve. Alternate versions include the Estaz Minnow on page 44.

Freshwater Flashabugger

This Flashabugger is an excellent go-to pattern in both fresh and brackish water—whether there's visible feeding or not. I prefer the black version in totally fresh water, especially when fished near emergent vegetation.
STEVE KANTNER

Hook:	#4 Mustad 9671
Thread:	Black 3/0 Monocord
Weight (optional):	.020 Lead wire or bead
Rib:	Black Kevlar
Weedguard:	10-pound Mason's Hard Mono
Tail:	Twin black marabou blood plumes with the stems removed.
Body:	Black olive Short-Flash Chenille
Hackle:	Extra-long dark dun or black genetic webby saddle.

Notes: I add a turn or two of chartreuse thread to the head if the fly is weighted. I like genetic saddles, grade #2 dark dun or black, that I tie in at the head and wind back over the body—before tying them off at the hook bend with the ribbing—which is then wound forward in a back-and-forth motion and tied off at the head. After securing the hackle, I bring the hard mono forward and tie it off at the hook eye.

Snakes Alive

When largemouths are celebrating their nuptials, they're less interested in food than in sex, but if you're willing to ignore the "do not disturb" signs, you'll get a rise. That's essentially what you're doing when you toss them the tidbit that I affectionately refer to as the Snakes Alive. It's my interpretation of a meddlesome creature— a pillaging water snake—that appears in their nest with larceny in mind. Snakes eat fish eggs, as well as fry, and this triggers angry responses from the largemouths, especially during the evening, when snakes go a' prowling. I wanted this pattern to combine action with realism—and yes, I was aiming for trophy catches. The Snake pushes water like its air-breathing namesake, while the rabbit-strip tail displays a menacing wriggle. That's usually enough to get a strike.

Bass typically spawn in secluded pockets isolated by heavy cover—inside a circle of grass, for example. So when you fish this pattern, add a stout mono tippet: 15- or 20-pound test. The pattern is most effective from November to April—when big bass are on the beds. Spawners prefer it ten to one over my other patterns.

Snakes Alive

This fly is deadly on bedding largemouths. Spit on your fingers and wet the wing before dressing the head with floatant. Otherwise, some dope will come off when it lands and prevent the fur from absorbing water, thereby ruining its action. STEVE KANTNER

Hook:	#1-1/0 Tiemco 8089
Thread:	Black flat-waxed
Tail Support:	Hard mono loop
Tail:	Black rabbit strip
Head:	Spun deer body hair
Eyes (optional):	Hollow plastic
Weedguard:	Double mono loops or wire

Notes: The rabbit strip tail measures 3 or 4 inches long and ¼ inch wide and is tied with the hair facing downward. I color the exposed skin with a black Magic Marker. The tail support should extend ¾ inch past the hook bend. If you use mono loops for the weedguard, bind them to the shank at the bend and then once again at the hook eye.

Legs Diamond

Fly fishermen haul untold numbers of largemouth bass from the Everglades and other freshwater haunts. Most are small, and if we ask the experts, it's because of the size of our lures. When it comes to fly casting, size equals difficulty. So we're mostly content to cast size 6 poppers and settle for those ¾- to 2-pound stud bass. If you'll settle for that, read no further. But if you've ever wanted to go for the gusto, I recommend the Legs Diamond. More so than oversize poppers or subsurface swim baits, it brings up the monsters while weeding out dinks.

More than one angler who examined my fly box commented on a slim hairbug that, in reality, is the size of a well-fed tarantula. I discovered rubber overkill in a photo of Kevin McEnerney's Crazy-legs Spider that appeared in Dick Stewart and Farrow Allen's excellent *Flies for Bass and Panfish* (1992). I immediately decided on a larger version, with modifications. I call it the Legs Diamond, at the suggestion of former *Warmwater Fly Fishing* editor John Likakis, who came up with the name.

While this hand-size monstrosity may resemble a fruit bat, its streamlined profile and non-absorbent materials make it easier to cast than most smaller hairbugs. The materials I use to tie this bug squeeze together when it's in the air and then splay when it lands. Since the legs absorbs little water (if it's properly greased), it's a snap to pick it up and false-cast. I can throw it for hours without fatigue. I use short, heavy leaders with heavier tippets—at least a 15-pound Maxima Chameleon. I also keep three or four fresh Legs on hand—dry bugs that I've greased with floatant—to switch out the waterlogged fly with a high-floating one.

I tie this pattern in both black and chartreuse, with legs to match. Lefty Kreh—who has caught practically everything that swims—suggested combining both the yellow and green shades in all-chartreuse lures. So far, it's working.

Legs Diamond

The Legs Diamond is my favorite pattern for supersize largemouths, especially after the water starts rising. When the Legs Diamond lands on the water, its footprint expands. Then, all those legs keep pumping throughout the retrieve. Together with the subtle wake, that opening and closing of the legs drives big bass nuts. It works extremely well for big fish in Everglades canals from late May through mid June—just when the water is starting to rise. STEVE KANTNER

Hook:	#1/0 Tiemco 8089
Thread:	Black Flymaster Plus or Flat-Waxed
Legs and Tail:	Solid-color rubber, three times body length
Body:	Spun deer body hair
Weedguard:	Mono loops or wire moustache

Finished Minnow

This fly rod plug excels during the first several weeks when the bass start to feed but still refuse to break the surface, due no doubt to their lowered metabolisms. The water's still cold. Yet you know the bass are eyeing the surface and gulping baitfish just beneath it. Since the bass won't break through the film, they won't hit a popper. This behavior persists until the water warms. It may take a few casts to find the right retrieve, but a simple bump-and-twitch usually works for me. The bass grab a Finished Minnow just under the surface, without much of a swirl.

While I am a big fan of weedguards, the Finished Minnow's crowning glory lies in its wobble—which would be severely hampered by extraneous fittings, including a wire or mono weedguard. So retrieve this lure alongside, rather than through, floating weeds.

Finished Minnow

My Finished Minnow is a long-rod rendition of the surface Rapala that dives and wiggles like all those highly effective twitch baits, which can be twitched and rested or steadily retrieved. Both imitate an injured baitfish The decal eyes and reflective sides resemble a balsa wobbler. STEVE KANTNER

Hook:	#6 Mustad 37187
Thread:	White Monocord
Body:	White foam pencil popper body. Spray the sides silver and the back blue, green, or brown.
Hook:	Long shank light wire, glued in place with five-minute epoxy
Flash:	Pearlescent Flashabou
Tail:	Olive marabou blood plumes

Notes: Cut the hard foam pencil popper body with a hacksaw so the face angles downward. Glue the body to the hook with five-minute epoxy. Then, after the epoxy sets, spray paint each side silver and the back green or gray. Glue several strips of Flashabou to each of the sides after catching a hank around the hook eye. Afterward, coat the body with clear acrylic.

Small Stuff

I designed the Small Stuff with small fish in mind. During a typical outing, I might release several dozen yearling bass and bream and maybe an oscar or two while twitching the Small Stuff. But since larger fish often target smaller baits, depending on where I'm fishing, it could be a succession of butterfly peacocks, along with some snook or freshwater tarpon.

Nonbelievers should try bouncing a Small Stuff near emergent vegetation whenever they see dragonflies in the air. Predators may think it resembles a nymph or even a grass shrimp. Strip it faster to imitate a baitfish, and you might be surprised at the results. Even at rest, the hackles keep breathing, and during retrieves, they practically come alive.

Small Stuff

The Small Stuff is my favorite subsurface panfish fly. I originally tied it to fool Mayan cichlids, but it has proven effective for a variety of species. This fly imitates baitfish, crustaceans, or nymphs: all common backwater forage. Vary your retrieve to imitate any of these food items. STEVE KANTNER

Hook:	#8 4 XL
Thread:	Olive 8/0 Uni-Thread
Weight:	Bead eye (tiny)
Rib:	Tan Kevlar
Hackle:	Two mottled olive turkey marabou plumes
Body:	Tan Ultra Chenille
Flash:	Pearl Flashabou

Notes: Tie in one of the hackles after wrapping part of the body. Tie in the flash at the eye and secure it at the hook bend with a Kevlar rib.

Veggie Flies

Urban targets for veggie flies, based on their availability as well as their willingness to hit a fly, include two native mullets, the lumbering Amur or Chinese grass carp, several varieties of tilapia, and, though uncommon, pacu—essentially a vegetarian piranha.

Mullet (*Mugil*) can live practically anywhere, from the ocean depths to freshwater swamps. They feed by sifting organic matter from ooze on the bottom or grabbing surface debris as it drifts with the current. Mullet are enigmas—to be present in such numbers, yet so seldom take a hook. Although incalculable numbers are netted for bait, only a precious few are caught on sporting gear. Fly fishermen occasionally land one as bycatch, and even fewer succumb to lures. Incidental hookups usually result from casting small streamers intended for other species. I remember as a kid watching subsistence fishermen pull mullet from a local canal. I marveled at how they did it: with bread balls and bobbers.

Since mullet are forage for larger fish, when they're in open water, they're either on the go or too nervous to feed. But locate these fish in a secluded canal, and the scenario changes. When mullet get a chance to relax, they start feeding again—typically on algae and other plant material. That's when sleepy backwater truly pays dividends.

Both black and silver mullet require a fair amount of stroking, or chum. Baiting the field with oatmeal or dry batter mix is how most folks get things started. Once mullet show up in the slick (you'll see the oatmeal flakes disappearing), ramp things up by tossing out bread crusts. That moves the action up on top. Part of the appeal comes from casting to schools that move furtively from crust to crust. If swirls appear, then I've hit the jackpot, so I let the fish start feeding in earnest while I toss out more bread. In time, the schools establish a pattern, which lets me lead them like a covey of game birds.

After casting my fly—a ball of Antron with or without a modified wing that's rigged below a strike indicator—I allow it to sink, until it's hanging approximately a foot beneath the surface. If it's properly positioned and a school of mullet swims past it, my "bobber" bobbles and, ultimately, dives. Getting a small white speck in front of a hungry mullet is what matters most. Dick Kirkpatrick, an old editor friend who loved to experiment, took a fair number of mullet on Light Cahill dry flies.

Yeastie-Beastie

The Yeastie-Beastie. Call it my all-purpose mullet fly. Antron sinks slowly and blends with the water: How's that for dissolving bread? Poly, on the other hand, floats—which can prove helpful late in the day. STEVE KANTNER

Hook:	#14 Mustad 94840
Thread:	White 8/0 Uni-Thread
Tail:	White Antron
Body:	White Antron
Wing:	White Antron

Notes: The body is made from white Antron wrapped back and forth in the shape of a football and tied off at the head. The tail is a tuft of fibers left over after tying off the body.

The Bisquick Nymph is similar to the Yeastie. However, polypropylene yarn (or dubbing) is substituted for the Antron.

Frank Oblak, an old friend from my pier fishing days, shares my penchant for quirky fishing. He, more than anyone, deserves credit for producing the first deliberate mullet flies here in the States. Frank carefully studied local bread-ball fishermen before fashioning patterns that looked and acted authentic. To the best of my knowledge, Frank was the first to deliberately target both the black and silver mullet after chumming the water with bread. All I did was modify his dressings by substituting Antron for polypropylene yarn after witnessing Antron's magical effect on freshwater trout. (Antron's index of refraction is similar to water's, so underwater, a ball of white Antron resembles dissolving bread.)

These patterns all serve a similar purpose. Any differences are due to sink rate and profile, with the heavier dressings (say, the MS7) reserved for larger mullet. And yes, you can see if any monsters are hitting the chum. As for sink rates, the action moves upward as the shadows lengthen. Colors, too, are interchangeable: tan fills the bill when rye's on the menu.

MS7

The MS7 is similar to the Yeastie, but the wing is longer. STEVE KANTNER

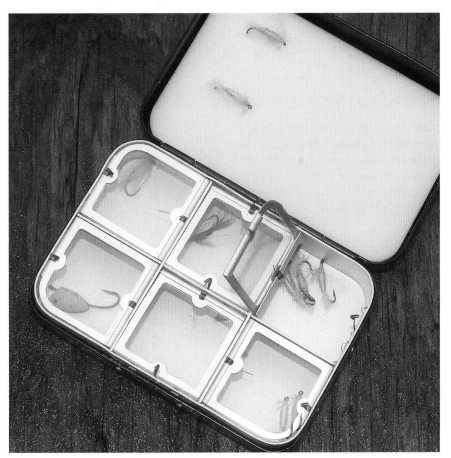

Hook:	#14 Mustad 94840
Thread:	White 8/0 Uni-Thread
Tail:	White Antron
Body:	White Antron
Wing:	White Antron

Notes: Wrap the yarn body in a football shape and tie it off behind the head. Leave a veil of Antron fibers slightly longer than the body for the wing. The wing is made up of these fibers left over after tying off the body pulled backwards. Leave the wing longer than on the Yeastie.

Since the mullet fishery is defined by uncomplicated tufts of fuzz that resemble dough balls, intricate patterns with carefully constructed legs and perfect hackles aren't part of the game. However, the right size fly is. I carry assorted patterns in various colors and sizes in order to match the chum. STEVE KANTNER

Several varieties of tilapia round out the vegetarian trifecta. If you've ever seen one in your neighborhood market, you know they resemble a bloated sunfish with bulging eyes. From an angler's perspective, they're skittish creatures reluctant to strike standard baits or lures.

Tilapia are raised commercially for food. Some authorities suspect that they infiltrated our waterways as a result of accidental fish farm releases. However it happened, they make up a substantial percentage of the biomass found in South Florida freshwater lakes and canals. Both spotted tilapia (*Tilapia mariae*) and the blue variety (*Oreochromis aureus*) are known to stray into brackish water. A third species, the Mozambique tilapia (*Oreochromis mossambicus*) is not as common and more difficult to catch. Because their feeding behavior is much like a mullet's, fishing for tilapia is similar, too.

There are two ways to catch tilapia. During their spring spawn, spotted tilapia will strike a #8-10 black Woolly Bugger that's repeatedly jigged near wherever they're bedding (which usually isn't far from where peacocks spawned earlier).

The second way to catch them is to chum for them with bread—just like you do for mullet, with a tiny fly and a strike indicator. Use flies tied with olive Antron, like an olive Yeastie or a Moss Fly.

Moss Fly

Think "gob of ooze" when you tie the Moss Fly for vegetarian tilapia. I tie tilapia flies with synthetics—mostly Antron or polypropylene, depending on whether I want the fly to sink or float. Here, as with mullet, my preference is for flat-forged trout hooks.

Hook: #14 Mustad 94840
Thread: Olive 8/0 Uni-Thread
Body: Green Antron
Wing: Green Antron

Notes: For the body and wing, sometimes I substitute polypropylene, depending on whether I want the fly to float or sink. Poly is more buoyant than Antron.

Stealth is essential when you're casting to grass carp, which is why long and fine leaders are so important (I want mine at least 10-feet long, and to taper to a 6 or 8 pound test tippet). Your fly should land like a snowflake. Afterwards, avoid unnecessary false casts and drag from any current.

Pacus, which hit bread balls and berry flies like grass carp, are occasionally found beneath ficus trees. However, hook a pacu and you'll see the difference—line melts off your reel like you snagged a freight train. One that lived in a tank at the Florida's Non-Native Lab's former location went by the name of Igor. Igor weighed at least 25 pounds. I've seen pacus in the wild that were even larger.

Original Kantnerberry

The original Kantnerberry—which imitates the fruits of bankside ficus trees—was designed for fooling persnickety grass carp. I tied the originals from old-style deer hair before switching to Otto's Inedible Egg. STEVE KANTNER

Hook: #8 Mustad 94840
Body: Cool purple deer body hair
Notes: Trim the body round with a flattened bottom leaving plenty of clearance. Cool red is another popular color, but grass carp run if the color's too bright. Go light on the lacquer and avoid using floatant. Carp have the nose of a hound.

Otto's Inedible Egg

Otto's Inedible Egg is a state-of-the-art temptation for grass carp. I named this pattern for Dr. Otto Lanz, a veterinary professor at Virginia Tech who helped invent it, and who was among the first to catch grass carp on it. He got corks from the chemistry lab, where they were going to waste. Now you can order cork balls online.

Hook: #4 Gamakatsu SC 15
Body: 10 mm cork ball, colored with Avon Cherries Jubilee nail polish
Notes: If you can't find cork balls, then try strike indicators—preferably cork—or chop up a wine cork and shape it with sandpaper. Make a slot in the cork ball and glue it to the shank with five-minute epoxy. I cut the slot with the edge of a hook file before gluing the cork to the hook shank without wrapping it first with tying thread. That makes for a stronger bond. Flattening the barb helps keep my berry flies intact. Plus, there's a lot less damage when I release a carp.

Lures for the In-Between

In this peripheral zone your patience will be tested—especially in waters that drain the salt marsh. Flies take top honors in mangrove-lined settings, where occasional traffic is the major distraction. But that doesn't mean that hardware won't work there, specifically tiny Rapalas and unweighted shads that are best cast on spin gear. While the fish are large, effective lures aren't.

If there's a cardinal rule, it's this: Fish feed up and not down, with the possible exceptions of sturgeons and stingrays—those two I'm not sure of. Honest-to-goodness surface feeding, where fish break the film, typically takes place under low-light conditions at dawn, at dusk, during stormy weather, or whenever the water is noticeably off-color.

In the suburbs, spinning is standard, and I make every effort to travel light. For roadside canals, I like fishing unweighted shads on offset hooks (Gamakatsu or Owner) or marabou jigs with lightweight spin gear. I carry a few plugs in my bag, as well as a box filled with Bass Assassins, three or four offset hooks, several quarter-ounce marabou jigs (the heaviest I can find), a No. 5 surface Rapala (black back/silver sides), and a Rapala Shad Rap or Rat-L-Trap.

If the water looks deep and I'm after tarpon, I'd start by casting a white crappie jig on ultralight line because of the light wire hook. I'll retrieve it a foot or less

Clockwise from top: Rapala floater, Zara Puppy, homemade popper, D.O.A. Terror-Eyz, Bill Dance topwater, 4M MirrOlure, 3½-inch Bass Assassin, marabou crappie jig. STEVE KANTNER

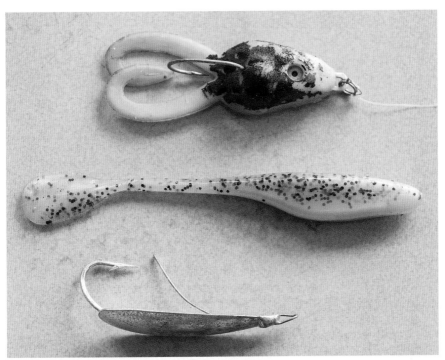

If I had three lures to wade with in a marsh, here's what they'd be: first, a 6-inch Gambler Flappin' Shad; second, a small, black Johnson Silver Minnow; and third, one of those new weedless frogs (like Bill Dance's Swimmin' Frog, the Cane Toad, or Zoom Horny Toad). The Flappin' Shad is probably the most versatile. Fish it with or without weight, with its nose bitten off as a popper, or tear off the paddle tail and use it as a shorter version. Plastics are inexpensive, and fit nicely in your pocket for easy access. Be careful of the hooks, though. STEVE KANTNER

beneath the surface while gently jiggling (not jigging) my rod tip. The jiggle is important, and most beginners get it wrong. You're trying to activate the marabou fibers by making them pulse. Jigging, on the other hand, involves jerks and pauses. If in doubt, simply reel the jig and allow the current to work the fibers. Either way, try to reel with the current—in direct opposition to how you fish a fly.

When I'm looking for peacocks, I'll fish a deep-running crankbait.

Most unweighted shads have a flattened appearance and sometimes a slit for the hook. You can either fish them bare or add them to a jighead. Some versions have tails that add to their action; those are usually attached to jigheads—typically for fishing the salt.

These lures mimic their namesake, the threadfin shad, but they also imitate guppies and mollies when fished with minimal weight. I'm partial to Bass Assassins in the 3½-inch length. My favorite Assassin colors are albino shad and rainbow trout, followed by gold flake/brown back. They're effective for snook and tarpon, as well as for snakeheads and sometimes peacocks. As you probably guessed, they're deadly on bass (without being lethal).

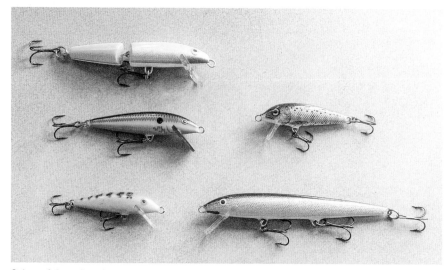

Colors of these Rapalas here range from "fire tiger" and chartreuse to silver/black back and "rainbow trout." They're great for tarpon in open water, but watch out for those trebles! Rapala rap sheet: sizes 3 through 7 lead the local hit list. STEVE KANTNER

Butterfly peacocks open new possibilities for anglers, while helping control Florida's less desirable exotics. These brilliantly colored gamefish are quick to strike lures: small ones, especially. Guide Alan Zaremba's favorite lures include floating Rapalas; suspending jerk baits like the Strike King Wild Shiner and Rapala Husky; surface plugs like the Heddon Torpedo, MirrOlure MirrO Prop, High Roller, Chug Roller; and Rat-L-Trap lipless crankbaits. PAT FORD

I rig these lures for the salt marsh like I would for largemouths, with the hook point barely exposed. If I need to fight the wind, I'll stuff a tiny, $\frac{1}{32}$-ounce weight in the nose. For snook or tarpon, I add a short piece of 50-pound green super-braid for leader (with a double uni knot) that allows freedom of movement while preventing cut-offs. (A heavy mono leader stifles the action when you're fishing something as light as an unweighted shad.)

My favorite peacock lure is the Rapala Shad Rap. It's a suspending plug, so you reel it down before working it back to the bank with a series of twitches. Admittedly, this lure sports multiple trebles, which can cause damage. But the larger the lure, the lower the chances of success. When I guided regularly, I used hookless plugs to fish during periods of high or off-color water. I'd cover as much water as I could, with my fly-fishing sport in tow. Whenever I detected a flash or a bump, I'd have my client cover the area thoroughly. It worked like a charm.

Flies for the In-Between

The salt marsh and suburbia are miles apart, but I include them together due to similarities in salinity content. Boundaries blur between brackish and interior habitats, as evinced by the species you'll find in each. While the following patterns perform best near the coast, they're also effective in freshwater ecosystems.

Salt Marsh Patterns

Flies are extremely effective in the salt marsh, and my essential patterns are Suescun's Conehead, the Rivet, Conner's Glades Minnow, Tabory's Squid Fly, the Flashabugger, various sliders and poppers, and Marabou Muddlers. I carry specialty patterns (shrimp imitations, mostly) for oddball species—sand perch and cichlids— that disregard my basic flies or for a client who makes a specific request. One fly fisher was determined to catch a mudfish, so out came a Clouser. You'll find times when mainstream species want offbeat patterns: Take when snook get the urge to feed on grass shrimp. I reserve a second group of flies for suburbia, including Clousers and berry flies.

You need to carry patterns in a variety of colors, patterns that ride in the water differently, patterns that are fully dressed for off-color water and sparsely dressed

Rivet rides rigged on a car top rack. Ease of tying, matched with all the right moves, made this fly my favorite for fishing the salt marsh. In a sense, it created a career. STEVE KANTNER

for clear water, drab flies and bright flies—all kinds of patterns for the fish's mood and water conditions.

Brightly colored flies work as an occasional alternative, but they're overkill except under certain conditions. Be mindful of size, and keep in mind that bigger isn't always better. A size 2 hook is usually my limit. Even largemouth bass, after entering the marsh, grow accustomed to tiny fare. A perfect example is how the bass prefer Mike Conner's Glades Minnow in size 6 or 8. Don't confuse these migrants with their Everglades cousins that prefer larger, more buggy lures.

Mike Connor's Glades Minnow

When it was first introduced more than a decade ago, the Glades Minnow was heralded as the quintessential Everglades fly. Today, this fly remains the pattern of choice for a hardhead imitation. Created by Stuart, Florida, angler Mike Conner in an attempt to imitate the tiny mollies and mosquitofish that infest these salt marsh canals—the Glades Minnow became a salt marsh staple. In addition to plaudits earned in the salt marsh, this pattern reaps praise from anglers worldwide.

One of the identifying features of this subsurface pattern is its expanded under-fur head. Flies that push water—mimicking hardhead baitfish—can make the difference between success and failure. To create this effect, Mike used a comb to collect acrylant fuzz from Kraft Fur backing before tying in small clumps at a time—like you would for wool-head flies. But unlike wool, Kraft Fur sheds water, making the Glades Minnow (first called the Trail Minnow) lighter and easier to cast.

If it's tied correctly, you won't see the thread. This seamless effect results from a careful melding of materials, monofilament, and head cement. I prefer mine in natural colors, tied on size 2 and 4 hooks, although I tie a "butter" yellow- and green-chartreuse version that's effective for backcountry tarpon under low-light conditions. You'll note that there are two shades of chartreuse. Green chartreuse has more of a lime coloration, while yellow chartreuse is more lemon colored. If you accept Conner's interpretation regarding the shape of a hardhead, how do you deal with the retrieve? The answer, I discovered, wasn't cut-and-dried.

One day early in my guiding career, I hooked several dozen snook in quick succession, and lost the majority to tackle failures. All, I believe, weighed at least 8 pounds apiece; a few looked better than 20. But since they were feeding on rain bait, I was forced to use a size 6 Glades Minnow, tied directly to my 12-pound-test tippet.

This took place within a few yards of the highway. I can still close my eyes and still hear the traffic tooling past me as I cast to repeated splashes. The melee—true mayhem for as long as it lasted—was centered around a single emptying creek. But try as I might, I couldn't buy a strike.

I saw breaking fish wherever I looked, and I kept casting feverishly in an effort to trick one. Yet none would oblige until I stumbled onto a technique that a freshwater trout fisherman would have gone to immediately: fishing the fly like a nymph.

I'd been casting to swirls and trying different retrieves when I paused to brush off a bug. That's when the first snook hit. Although the hook tore out an instant later, the same thing happened on a subsequent cast—once again, when I stopped my retrieve.

No matter how lifelike my fly appeared, these snook weren't interested in chasing my inch-long baitfish. They were gulping micro-forage like trout atop weed beds slurping captive emergers. When I simply let it sink, watching for the line to twitch, I could barely believe the result.

A snook inhaled the fly before it neared the bottom, heralding a succession of similar responses. Whenever my line twitched, I'd come tight to a fish that took off with me in hot pursuit. I spent the next two hours dodging cattails instead of tearing my hair out.

Glades Minnow

The original Glades Minnow remains my go-to pattern wherever overstuffed salt marsh snook demand exact imitations. Mike initially tied his signature pattern to tempt the snook and tarpon of the Tamiami Trail, but it works in all South Florida backwater for a variety of species. This one was tied by Conner himself.

STEVE KANTNER

Hook:	#4 Mustad 34007
Thread:	Clear monofilament (fine)
Body:	Gold Mylar braid
Wings:	Light over dark, twice the shank length, with a pinch of superfine Angle Hair in between. For color, I prefer medium brown over tan, or green chartreuse over yellow chartreuse.
Flash:	Gold Angel Hair
Head:	Acrylant fuzz, picked from a comb after running it through Kraft Fur.
Eyes:	Reflective gold or red decal

Notes: I coat the thread wraps with superglue before adding the body. Wind around the shank for the body.

For durability, place a drop of five-minute epoxy over each eye after completing. You can use a drop of superglue for head cement after each tying step.

Alex Suescun's Conehead

I've seen the term "conehead" listed in a retailer's catalog, but the fly above it looked nothing like Alex's. Like its inspiration, the Creek Chub Darter, he designed it to swim with a side-to-side motion that drives backcountry predators wild. By combining super-soft wings and a diaphanous collar, and adding a specially trimmed deer hair head, Alex created a combined imitator and attractor pattern that's effective wherever you fish—not just in the salt marsh.

The Conehead gets its name and its signature action from the shaven deer hair head. The action is so pronounced that it rivals its wooden antecedent. Suescun's design has the pop and sizzle of a traditional streamer with its saddle hackle wings, rabbit fur or marabou collar, and deer hair head.

Pink Suescun's Conehead. When it comes to waylaying trophy snook, many veterans rely on time-tested plugs. That's what Alex Suescun had in mind when he created his iconic Conehead streamer, easily the most effective backwater fly, especially for tarpon. STEVE KANTNER

The result is a pattern that literally cuts through the water, while displaying maximum bounce. Watch a Conehead in action and you'll see what I mean: a spectacle that's seldom lost on the fish. The deer hair head, which is naturally buoyant, allows you to fish this fly as a waking pattern. Depending on how big the head is—and whether or not you grease it—the Conehead can be twitched as a slider or retrieved just under the surface. The dressing can be modified to suit your specific requirements. My salt marsh friends like theirs tied in white, black, or yellow—with grizzly outer wings—and they fish them as splashy attractors. They also fish neutral versions that may imitate a baitfish.

The Conehead looks enough like a pike killifish to qualify as a match. Pike killifish grow larger than guppies or mollies, so the hook, for a change, needn't be so tiny.

I tie Coneheads in brighter colors (chartreuse or pink) for murky water. Be careful, however, to avoid being too vivid. The Conehead excels when predators aren't keying on anything special. Since Coneheads are attractors, I don't worry as much about matching sizes as I do with the Glades Minnow. I tie them on hooks up to size 1.

Tarpon, South Florida's largest backwater fish, prefer size 4 or 6. With a fly that small, you're consigned to lighter tippets, so the fish either frays through the mono or the hook tears out due to its narrow gape after a protracted battle. Use anything larger and the fish ignore it. But not so with Coneheads, which both serve as attractors and imitate larger fare.

The Conehead is my number-one choice for backcountry tarpon. Snook like it, too—especially the big fish that swim up from the Gulf during the fall run. They're accustomed to larger fare, so I don't waste time with tiny offerings. This is one of the few times when bigger is better. I use a white, pink, or chartreuse Conehead (depending on water clarity) tied on a size 1 hook.

Conehead

Fully dressed white Conehead (side view). A few of my friends—skiff fishermen, mostly—tie Coneheads with oversize heads strictly for fishing on top. I prefer a tight, lacquered head and sparser dressing that rides just beneath the surface, where the fly creates a tiny bulge.
STEVE KANTNER

Hook:	#2-4 Mustad 34007
Thread:	210 Denier Danville FlyMaster Plus
Wing:	Six white saddles per side, each set bordered by a thin grizzly saddle.
Flash:	Flashabou
Collar:	White marabou blood plumes
Head:	Spun deer body hair

Notes: I glue the wing butts together before tying in each side. For the head, pack the hair hard and then trim it with a razor (be careful!). The head should appear triangular when viewed from the front.

Rivet

When I first ventured into the mangrove backcountry, I had no idea where to start. Should I imitate forage or fish attractor patterns for the saltwater predators that plied these creeks? Although I saw visible feeding, I couldn't always get strikes. Snook frequently ignored the flamboyant Conehead, especially when feeding on tiny rain bait. Then when I imitated rain bait with Connor's Glades Minnow, the fish might not hit it or the hooks pulled out. I needed to make them an offer they couldn't refuse.

Enter the Rivet. Its color—or lack of it—is accurate, and its translucence and side-to-side wobble brought snook running. Plus, I could tie it in larger sizes and still get plenty of hits.

While the Rivet resembles a Conehead—at least, to the untrained eye—the similarity ends in the hackle wings, the diaphanous collar, and the deer hair head. The Rivet wobbles less than the Conehead—because it has a stiffer rooster hackle or a Kraft Fur wing—and it derives sex appeal from its heavily dressed collar—actually two collars in one. Displacement is king in these brackish marshes. Since you're frequently fishing in off-color water, flies that create shock waves typically draw more strikes.

The Rivet is an attractor fly, but it also imitates the vaunted pike killifish. This bite-size tidbit reaches 6 inches or more in length and looks like a baby northern pike. But it's the piece de resistance of salt marsh forage. Snook, in fact, love killies so much that if they discover even a few, they forget about rain bait and pull out all the stops to focus on them.

Pike killifish invariably swim near the top, and between short bursts of speed, they curl their tails. It's this signature motion that predators look for. If the water's off-color and killifish are present, it's time to haul out a Rivet. For best results, I retrieve the Rivet in short, snappy strips with distinct pauses between them.

Rivet

Sturdy construction made the Rivet the sensible choice for constant guiding in the mangrove morass, where the shelf life of flies is measured in minutes. After tying the original, I came up with a second version that substitutes Kraft Fur for the neck hackle wings. This fly is excellent for snook in off-color water—in canals, especially. STEVE KANTNER

Hook:	#4 Mustad 34007 stainless
Thread:	Gray 3/0 Monocord
Wing and Tail:	Tan Craft Fur tied in at the hook bend
Flash:	Pearl Angel Hair
Collar 1:	White rabbit body fur
Collar 2:	Natural gray and brown rabbit fur
Head:	Gray spun deer hair

Notes: I use a felt-tipped pen to add several stripes to the wing and tail. The first collar, the white rabbit body fur, is half as long as the tail. Tie it in at the middle of the shank. The second collar should not be quite as long as the first. Imagine a lady's petticoat under a hoop skirt. The head should be trimmed fairly small.

Lou Tabory's Squid Fly (Modified)

There's a saying along the Tamiami Trail that the tarpon will hit only two fly patterns: black ones and white ones. Try as I might, I can't debunk it. Of all the backwater tarpon flies I know and love, the Squid Fly excels on two points—it's both effective and easy to tie.

Designed by Northeastern surf guru Lou Tabory, the Squid Fly lives up to its name. On fast-strip retrieves, the herl comes alive, making it resemble its namesake forage. But you won't find squid in the swamps. It must be some sort of sex appeal—call it bounce or flourish—that these salt marsh predators can't resist. Work this fly with twitches and pauses.

The monster stripers that Lou Tabory beaches would sneeze at my miniscule morsel. However, after adjusting the proportions to match the forage, I ended up with a killer fly. So what does it imitate? Who knows for sure? I'd say an injured baitfish. I typically tie it with a bicolor head. You can take my word that chartreuse attracts tarpon, although I can't remember why I started adding it to begin with.

Squid Fly (Modified)

Tabory's Squid Fly. Effective for everything, especially tarpon. Keep the ostrich wing short.
STEVE KANTNER

Hook:	#4 Mustad 34007
Thread:	White monochord
Wing:	White ostrich herl (from the tip of the feather)
Collar:	White marabou or rabbit fur, half the wing length
Head:	White spun deer hair, with a pinch of chartreuse added for color

Shiny Buggers

When I don't know what the fish are feeding on, I often use a generic Flashabugger pattern. Backwater tarpon strike Flashabuggers with abandon, as do snook and butterfly peacocks, which inhabit the same basic region now that peacocks were discovered just north of the salt marsh. Buggers shine on slick-calm days when the tarpon are listless. Although the fish refuse your other offerings, they'll often change their tune at the drop of a Flashabugger and chase it all the way to the bank.

A shiny black Bugger, possibly the best damselfly and dragonfly imitation, is more at home in freshwater haunts. However, my favorite for the salt marsh and other in-between venues is a version tied with purple chenille.

I keep two fly rods rigged when I'm pursuing peacocks: one with a black (or purple) Bugger; the other with chartreuse—preferably a combination of yellow and green shades, a trick I learned from Lefty Kreh.

I fish the Estaz Minnow for spawning peacocks to change things up. This fly has all the hallmarks of its more elaborate cousin the Flashabugger but takes less time to tie. How much less? Imagine a marabou tail and an Estaz body, wrapped on a hook shank with added weight up front. I use a tiny version when I fish the dock lights.

The Estaz Minnow (bottom) is a simple pattern that I use when peacocks are spawning. Above it is a Krystal Flashabugger. STEVE KANTNER

Flashabugger

You may recognize this pattern from the Freshwater Flies section (page 23). This versatile fly works in a variety of situations. STEVE KANTNER

Hook:	Mustad 9672
Thread:	Black 3/0 Monocord
Tail:	Purple marabou blood plumes with the stems removed, plus a few strands of Krystal Flash
Rib:	Black Kevlar
Body:	Purple Ultra Chenille
Hackle:	Long webby purple saddle
Weight (optional):	Lead wire or bead

Notes: I tie the Krystal Flash strands at the head and eventually pull them over the body, before securing them at the hook bend with the Kevlar rib. I add a turn or two of chartreuse thread to the head if the fly is weighted. After tossing the fly out and allowing it to sink, I retrieve it slowly while twitching it gently. It's just the ticket for lazy tarpon. Meanwhile, the black version, which also resembles certain salt marsh baitfish, works well under bridges for snook.

Poppers and Sliders

Both the noisy poppers and their streamlined cousins, sliders, work well in the tangled warrens where the Glades meet the Gulf. The rules that govern them seldom waver, nor does the sequence in which each becomes the go-to lure. Remember, gamefish feed closer to the surface as they lose their caution. However, when it comes to making the ultimate commitment, fish, like humans, prefer to test the waters. Just luring a fish that final half inch is tougher than drawing it several feet from the bottom. There's a world of difference between exploding on top and breaking the film. Forget what happens once you hook a tarpon; fish, as a rule, aren't partial to air. There are times, however, when they're focused on the surface and turn up their noses at subsurface offerings.

The slightest variation in where a fly rides may determine whether predators will hit or refuse it. But keep this in mind: Just because they'll hit a slider doesn't mean that they'll hit a popper—or vice versa.

Fish scan the surface during low-light periods at dawn or at dusk or on cloudy days. Off-color water limits the amount of light that can penetrate the surface too. From beneath, the surface is a shimmering mirror against which surface lures show up in bold relief. Too much disturbance and fish head for the hills (or channels); too little and they ignore whatever you're selling. So how can you stay on track? By being conservative and moving slowly.

The legendary Captain Bill Curtis and the 27-pound snook he caught on a slider, while fishing the canal on Highway 29. I know that spot, but I've never been able to repeat Bill's performance.
BILL CURTIS VIA PAT FORD

Foam-headed pusher bugs evoke powerful strikes from bullseye snakeheads. STEVE KANTNER

If fish will hit on top, you'll see multiple clues. Foremost among them are repeated pops and swirls from which leaping forage attempts to escape. The next are muddy water or low-light conditions. If the water's shallow, that's a good indication, since the fish don't have far to rise. Combine any two and you could be in business; all three and you're ready to launch.

That's when the snook and tarpon go on a rampage and demand a topwater bait.

Here's typically how it goes: you're getting frequent strikes on a subsurface pattern when loud pops start erupting in the waters around you. Then, all of a sudden, the fish quit hitting. You notice their attacks have intensified, but you'd hardly know it from the results you're getting. So how can you turn the tide?

Your luck will probably rebound if you switch to a waker like a Marabou Muddler or a well-greased Conehead—something that, for the most part, remains in the film. Then, as caution devolves into reckless abandon, what worked before is suddenly passé. That's when snook or tarpon look for something splashy: a slider or popper, usually in that order.

When the fish completely lose their inhibitions, they'll smash a popper on every cast. However, those leg-studded monstrosities that shine in the swamps have limited value here—even to salt marsh bass. The fish seldom require more than a slider. And if they do, it's not for long.

I've had marginal success with the foam-disk versions of poppers, but I prefer tiny pencil poppers with hard-foam bodies: the ones you make from kits from Wapsi or Bass Pro Shops. Since the hooks that come with them are a bit too light, I replace them with long-shank Mustads (34011), usually in the next larger size, after sanding the slots to accommodate the shank. Another trick I learned is to heat the hook shank with a cigar lighter before forcing it into the slot. Just be careful not to burn your fingers.

All-Purpose Slider

Poppers and sliders excel in shallow run outs. I fish a slider such as this one before switching to a popper. STEVE KANTNER

Hook:	#4 Mustad 34007
Thread:	White 6/0 Flymaster or flat-waxed nylon
Body:	No. 6 slotted, tapered cork
Tail:	One or two white marabou blood plumes

Notes: I use the same corks for tying bass bugs but tie them backward on the hook shank for the slider. I paint mine white with Gesso before taking time to add the tail. The tail should be slightly longer than the body. I sometimes include a few strands of flash.

All-Purpose Pencil Popper

Pencil poppers like the one in this photo can be deadly on snook and tarpon under the right conditions. I've noticed that salt marsh fish that hit poppers run a bit larger than usual.
STEVE KANTNER

Hook:	#4 Mustad 34011
Body:	Slotted, pre-formed hard-cell foam, the thinnest you can find
Thread:	White Flymaster or flat-waxed nylon
Tail:	White bucktail (fine)

Notes: Don't wrap the hook shank with thread. Either epoxy the hook into the slot, after widening the gap with medium-grit sandpaper, or heat the hook with a cigar lighter (while holding it with pliers) and force the red-hot metal into the groove. Even the bucktail tips in a stacker and keep the tail extremely short. Use thread to attach the tail.

Black Marabou Muddler

One of the most effective patterns to ever drop from a fly vise is the Marabou Muddler. A descendant of Don Gapen's original Muddler Minnow, the marabou version resembles a variety of forage, as evinced by its widespread appeal. The secret's in the action, which includes this pattern's ability to bulge or wake because of its deer hair head. That, combined with marabou, creates the illusion of a *Gambusia* or mollie fleeing the scene.

Black Marabou Muddler

This one's a keeper wherever you fish. This pattern can be tied in any color on any size hook. Black Marabou Muddlers are effective in the salt marsh as well as below spillways. Yellow and gray are salt marsh favorites, while black excels below suburban spillways. STEVE KANTNER

Hook:	#2-8 Mustad 9671
Thread:	Black flat-waxed nylon
Body:	Gold Flashabou
Tail:	Fluff from the base of a red hackle quill
Wing:	Two black marabou blood plumes
Flash:	Angel Hair
Head:	Black deer body hair (fine)

Note: I use a trout hook because the flat-forged Mustad goes in easy like a needle, and I can resharpen it whenever I need to. It resists straightening or breaking, too, as few of its designer cousins can claim. I add black deer hair or squirrel tail if I'm adding an underwing.

Saltwater Lures

Surface Lures

Surface feeding is either one of two types: focused and deliberate or helter-skelter. Take cagey snook and tarpon, which are cautious feeders that seldom forage on top without provocation. An abundance of prey may be the catalyst—or darkness or competition from other predators. Even then, they examine their prey, if only for a second. So for solitary species, I employ a slower retrieve and make sure that my lure resembles their prey and that how I work it completes the illusion.

The helter-skelter feeding comes in waves—schools of ravenous, blood-glutted predators that eat whatever gets in their way, if it looks at all like what they're accustomed to chasing. Think bluefish, sharks, jacks, and false albacore (or bonito)—species that remain on the go, which builds up a caloric debt, meaning they're usually hungry.

Shoals of big jacks arrive during spring. Whether you're wading a grass flat or fishing the beach—or a pier, for that matter—you'll run into gamefish feeding on top. The feeding may look haphazard, like a slurping seatrout, or be visibly violent, like jacks or bluefish herding mullet. At times like these, when predators pull out the stops, it's still important to feed them properly. PAT FORD

For these eager beavers, I invariably prefer a surface lure—let's call them plugs, since hard baits excel at this type of fishing—with certain characteristics, like the right silhouette and at least one set of strong trebles.

I'm not a big fan of cup-faced poppers; they're too difficult to cast in the wind. And this is, after all, a shore-fishing book. Plus, I want something that looks like an indigenous baitfish: specifically, a mullet, ballyhoo, or flying fish (yes, they come close to shore here). That leaves me looking for a plug that runs straight: a Rapala Skitter Walk, a MirrOlure Top Dog, an Excalibur Super Spook (or even an old-style Zara), or a Lazer-Eye pencil popper. (It's not much of a popper, if the truth be known.) I buy Lazer-Eye poppers at Bass Pro Shops. If I really need distance, I can fill them with BBs.

How do I work these straight-running surface plugs? By casting and reeling fast enough to keep them on top. No jigging or hesitation allowed here. And if baitfish are running, I cast beyond them and then reel it back at breakneck speed. Another method is to blind-cast in the surf if big fish are running. The secret to the latter— and it makes a difference—is for the plug to surf down the face of a wave, instead of vanishing behind every swell. It's all a matter of speed and timing to make sure the lure looks natural and gamefish can see it.

Now, for tire-kickers like snook and tarpon (and sometimes seatrout): I use the same plugs but go to great lengths to match what they're hitting—which some-times involves a smaller version or changing colors—although here in the salt, color's not all-important. I don't want any BBs at a time like this. What I want is a lure that sits on its haunches until I make it bobble by twitching my rod tip, like walking the dog.

Subsurface Lures

If there's an all-around lure for saltwater fishing, it must be the jig. In sizes ranging from tiny crappie jigs to 4-ounce bombs, jigs have proven their worth wherever they're fished—from residential canals to the open sea. Perhaps the most popular version is an ordinary nylon-skirted lead head that performs yeoman service on sea-sonal run fish.

The large hackle jigs some anglers call "chicken feathers" fool snook and tarpon in the surf and on bridges, and on area piers they'll fool cobia too. Other jigs that are popular are lightweight white bucktails and the Red-Tailed Hawk—a white nylon jig with trailing red fibers. I reel both diagonally through the breakers. To a snook, the hawk—or a 6-inch chicken feather—resembles a whiting. Next in impor-tance, at least to surf and pier fishermen, are wobbling spoons like the Krocodile and Kastmaster.

Spoons and jigs are probably responsible for capturing more run fish than all other lures combined. The most effective way to fish them is to cast upcurrent and slightly into the wind and allow your lure to sink to the bottom (be careful if you're fishing over rock or reef) before beginning your retrieve—either straight-reeling with a spoon or jigging with a jig.

Most of the lures I fish in the ocean are hard baits, but for seatrout and redfish, species that feed heavily on crustaceans, D.O.A. Lures in Stuart, Florida, makes an assortment of soft plastics that really do the job. My favorite of the bunch is the D.O.A. Shrimp, followed by the Terror-Eyze and Finger Mullet. The larger models excel for snook and large, gator seatrout. The D.O.A. Deadly Combo combines the shrimp with a rattling float. Simply toss it out in waist-deep water and every so often give the float a pop. D.O.A. LURES

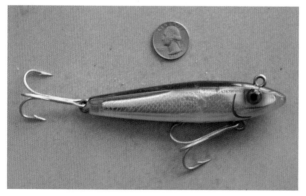

L&S MirrOlures have been popular for as long as I can remember. This particular model—the 85M18, which was discontinued—was deadly on cobia, tarpon and kings on the piers. Here's hoping the company brings it back. STEVE KANTNER

A variation of the spoon is the diamond jig, which is a cross between a wobbler and a jig and combines the advantages of both. (For one thing, it casts better.)

As far as finishes, silver is popular, although anglers nowadays like a splash of chartreuse. Wading anglers prefer gold for redfish.

Wobblers, the class of plugs that swim when retrieved, range in size from tiny Rapalas to larger Yo-Zuris and giant Rapalas. In the smaller sizes—say, a 10—X-Raps are deadly for inshore panfish such as Spanish mackerel. But another favorite model is the X-Rap 14, which is deadly on kingfish and will tempt tarpon. Before meeting the X-Raps, I enjoyed considerable success with wobbling Yo-Zuris—specifically the Crystal Minnow.

I've fished with numerous other models over the years, from Creek Chub Pikies to Rebels. Each has an action that's totally unique, and all are effective in the right situation.

Take the Pikie, which is an outstanding goggle-eye and mullet imitation, while the Rapala X-Rap 14 swims like a wounded sardine. I stay current with trends, but I never lose sight of the past. A perfect example is the MirrOlure 85M-18 (green back/silver flash), which proved its worth for over 40 years against a bevy of behemoths, from giant beachcomber kingfish and 100-pound tarpon, to outsized cobia and snook. Hopefully, the company will bring this discontinued model back.

In the salt, matching the hatch involves just a few species that look basically similar (known as whitebait) in addition to bay anchovies, mullet, and shrimp. PAT FORD

Saltwater Fly Patterns

Most saltwater baitfish, which these patterns represent, have silver sides and a gray or green back. Their major difference is their silhouettes, which tiers can shape to fit their needs. Saltwater feeding is usually spur-of-the-moment, but that's not to say that fish don't make distinctions in size or color. Bluefish, for example, prefer something meaty—say, a size 3/0 streamer tied from synthetic materials or, when conditions are right, a saltwater popper.

Fish need to see your fly before they'll hit it. A short shock leader fashioned from flexible wire helps prevent cutoffs.

My favorite patterns are simple yet durable ties such as Clouser Minnows, Lefty's Deceivers, and Brooks Blondes. Brooks Blondes are oldies but goodies, still effective whenever you need a whitebait imitation such as pilchard or greenie (thread herring). All-white and all-yellow are my favorite colors. I've caught hundreds of tarpon on the white bucktail version around Port Everglades Inlet when the water was clear.

Deceivers are my go-to pattern when I want to imitate something with heft. Tied in larger sizes (1 through 4/0) they are good imitations of gizzard shad, small blue runners, or adult thread herring.

Most of these flies are well-documented standards for the salt, so I don't include recipes or tying instructions.

A Clouser is a formidable weapon against snook in the surf and pompano in both the St. Lucie and Indian Rivers. You can tie it with natural or synthetic materials, heavy or sparse, to imitate a wide range of baitfish.

Lefty's Deceiver is effective wherever gamefish feed on other finny forage. Dressed sparsely, the Deceiver resembles glass minnows (bay anchovies).

This sparsely-tied Deceiver has caught plenty of snook.
STEVE KANTNER

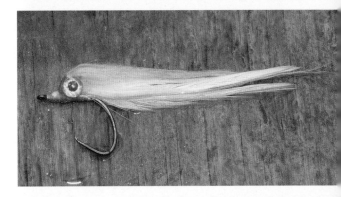

Another Deceiver (complete with plastic eyes) tied on an Owner Mutu Light offset circle hook.
STEVE KANTNER

This cigar minnow imitation is typical of the sparsely dressed streamers used in the salt.
STEVE KANTNER

Fully dressed Deceiver variant,
deadly for snook in the surf.
STEVE KANTNER

Another glass minnow pattern.
The trick is to tie them on long-
shank hooks (such as a stainless
size 4 Mustad 34011) while
limiting the amount of dressing.
These sparsely tied flies—most
include a strip of flash and a silver
body—imitate bay anchovies or
hog-mouth fry. STEVE KANTNER

Typical pilchard (or glass minnow)
imitation. Synthetics make this
fly more durable, an asset when
targeting sharp-toothed mackerel
and bluefish. STEVE KANTNER

I imitate glassies with a tiny Estaz Minnow—an abbreviated version of the Flashabugger—tied with a white marabou plume and pearlescent Estaz, with the body trimmed. Whether you call it Minnow or Schminnow, this pattern is effective wherever gamefish feed on rain bait. That's especially true in the Intracoastal Waterway under the dock lights. A trimmed Estaz-wrapped pattern, in sizes 6 and 8, is deadly for snook in the dock lights. STEVE KANTNER

Typical wool-head streamer for when the mullet are running. I use the same technique to fashion the head as Mike Conner did when he created his Glades Minnow, but I do it with wool. The wool and rabbit fur collar absorb water. Between that and the dumbbell eyes, this fly can become heavy. STEVE KANTNER

Saltwater poppers, with or without a cupped concave face, are effective for bluefish, jacks, or any predators that crash through baitfish. White with a red head or a silver body and green back are my favorite colors. STEVE KANTNER

The Interior

PART II

Florida's interior encompasses several million acres of amphibious prairie, pine and cypress forests, and weed-choked marshes. For the purpose of this book, it includes all the territory west of U.S. Highway 27, from Lake Okeechobee southward, and east of an imaginary line running from the southern tip of Big Lake O to Golden Gate Estates, a community that's located just east of Naples. The hammocks and grasslands lying north of that portion of U.S. 41 (the Tamiami Trail) that connect tiny Ochopee with Naples are also part of the interior. The whole interior is freshwater habitat.

Joannie's Crab House on U.S. 41. STEVE KANTNER

East of Ochopee, this region extends to within a few miles of Florida Bay, while north of U.S. 41 it includes all of the Everglades, the Conservation Areas, Big Cypress Preserve, and Corkscrew Swamp. Sections of this region stay dry all year; others remain submerged. The entire area is riddled with flooded grasslands or swamps that are connected by strips of land where hardscrabble pine and scrub palmetto grow. I include all the lakes and

Marjorie Stoneman Douglas called the Everglades the River of Grass. This photo was taken just west of Miami. STEVE KANTNER

Largemouth bass top the hit parade in Florida fresh waters. When concentrated in the Everglades by dropping water levels, they provide all-day action—especially on top. STEVE KANTNER

Aside from largemouths, you'll find an amazing array of other species that run the gamut from tarpon and snook to pickerel and panfish in our local fresh waters. The grass carp remains a favorite of mine, although I take extra care to release them unharmed. It is, after all, the law. PAT FORD

canals that lie at least partially within the densely populated strip of real estate that extends along South Florida's Gold and Treasure Coasts (see Part IV: The Salt).

The miles of flooded grasses that make up the Everglades are interspersed by a few man-made canals, the banks of which are lined by walls of cattails, making fishing extremely challenging. While the techniques used here are fairly straightforward, gaining access is not. Portions of this region are accessible on foot, but to enjoy its full bounty, I recommend a skiff or canoe, but not a kayak or float tube because of gators and snakes.

Timing

Fish in the interior follow a timeline that's based predominantly on water levels. Nine times out of ten, these water levels follow a schedule. When dropping water forces bass, panfish, and other species off the flats and into the grass-lined canals, they must compete. They'll lose all caution and feed all day long, and hundred-fish outings aren't at all uncommon. For the lowest possible water, I first look quite a ways inland, since this has a direct bearing on where the fish first go bonkers. The blitz begins in Palm Beach County before working south to Broward and Miami-Dade Counties.

South Florida's drought extends from just before Thanksgiving until around the third week in May. The summer rains subside in fall, and water levels start dropping by late December and continue receding until some former honey holes bake in the sun. Falling water forces fish off the flats—just as it does in the salt marsh—and into the main canals. When fish concentrate and start competing for food, fishing can be off the charts.

Everglades fishing used to be over by July. Now, thanks to the South Florida Water Management District's unpredictable pumping schedule, timetables are no

Diana Rudolph with a roadside snook from the Tamiami Canal. Fly fishing in the interior or in-between is just as effective as either spin or plug casting, yet it is not as popular as fly fishing in the salt. STEVE KANTNER

South Florida provides great fishing for a wide assortment of panfish, including the warmouth, the black crappie (sometimes called the speckled perch or speck), both bluegill and redear sunfish (or shellcracker), and "stump-knockers," or spotted sunfish. Here Martini Arostegui hefts a trophy bluegill from a residential canal, where they're a lot less common than redear sunfish or our myriad exotics. MARTY AROSTEGUI

longer etched in stone. If it's been raining profusely, the swamps will be flooded, and large volumes of water will head south. If it's dry as a bone, then the swamps may be fishable well into summer. Lesser bites may occur whenever the South Florida Water Management District runs water off the flats by pumping it southward through Everglades canals. Check the state's website to stay informed regarding pumping activity.

Most Glades canals appear on Water Management maps, like the one on page 67. Assume that the majority are lined with cattails, and that most support robust populations of gators and snakes. Gators, while torpid, will attack humans—mostly during their spring walkabout season. That's enough to discourage any would-be float tubers.

In addition to bass, you can find a wide range of panfish as well as native catfish—white, channel, and bullhead—in urban impoundments as well as swampy canals. Most panfish prefer natural baits, with night crawlers (for shellcrackers) and crickets (for bluegills) being most effective. Specks, on the other hand, prefer tiny minnows. Tiny jigs and grubs and fly rod poppers for bluegills and stumpknockers are also effective.

Where I live in Broward County, I expect good Glades action by the beginning of March. Bluegills start spawning then, too, during and immediately following full and new moons. It's a process they'll repeat throughout the summer, but in March they go hog wild. Then, a few weeks later, in April and May, big shellcrackers go on the beds.

Essential Gear

Today's typical bass outfit consists of a 6- to $7\frac{1}{2}$-foot medium- to heavy-action casting rod combined with a free-spool, bait-casting reel. Plug reels can be round or sit low on the reel seat. Line tests vary according to requirements, with super-braid performing better in heavy cover.

These gel-spun braids, which are manufactured according to one of two proprietary processes, are stronger, thinner, and more abrasion-resistant than mono—and they test nearly three times the strength at the equivalent diameter. Braids, however, can produce horrendous backlashes that only keen vision and patience can cure. These lines require special scissors to cut. With super-braid it's customary to add a short piece of 20- to 40-pound mono or fluorocarbon with a double uni knot.

For spinning gear, look for a rod with lighter action and a spinning reel capable of holding 150 yards of 12-pound-test mono. Ultralight spin tackle—4- or 6-pound-test and a $5\frac{1}{2}$-foot rod—is ideal for panfish and for casting tiny lures to bass in open water.

An angler hoists a chain pickerel that attacked a crankbait. Chain pickerel frequent canals that access marshes, especially canals with moving current. Pickerel hit spoons, as well as plastics and streamers, but nothing outperforms a fast-moving crankbait.
MARTY AROSTEGUI

This bass took a popper. By overlining your rod, you can make casts more efficiently with a short amount of line (less than 30 feet). Try fishing a line that is one or two weights heavier than what your rod is rated for. Specially made bass-bug fly lines are designed for this work. A short leader with a heavy butt helps turn over wind-resistant flies. For leaders, I prefer ones around 7 feet long that taper from a 30- or 40-pound-test butt section to a 6- or 8-pound-test tippet. STEVE KANTNER

I carry leader in sizes 30, 20, 10, 8, and 6—either Maxima Chameleon or plain, clear Ande. You'll also need floatant (for dressing the Legs Diamond and other hairbugs) as well as a fly box that's capable of holding a few basic patterns. STEVE KANTNER

Fly Tackle

A good freshwater fly rod and reel needn't be expensive, only functional. A reel with a click drag—just enough to prevent overruns—should suffice in the swamps, even if you stray into brackish water. While backing is optional, at least 100 yards of 20-pound Cortland Micron, or a suitable substitute, makes recovering line easier.

For tarpon up to 30 pounds, a 7-weight is ideal. However, certain situations—like going after bedding bass in heavy cover—call for something heavier. That's one place where a 9- or 10-weight isn't overkill. Casting large, wind-resistant lures puts power at a premium, so I up the bar to an overlined 9-weight, although I switch back to my 7½-footer for everyday fishing.

Dress for Safety

You may end up tromping where no wise man treads. But if you choose to chance it, wear protective clothing (including snake boots), and learn the location of hospitals in your area that carry a supply of antivenin.

Broward General in Fort Lauderdale and Jackson Memorial in Miami come immediately to mind, as does Naples General. Check for any recent updates. Incidentally, if an emergency physician has questions about treating snake bites, have them contact the 24-hour venom hotline at the University of Florida/Shands Hospital in Gainesville. While a fully-loaded hit from a coral snake or eastern diamondback has lethal potential, remaining calm and applying a tourniquet will tip the odds in your favor. Still, a snake bite is serious business.

Diamondback rattler. Numbers of snake are generally on the decline. Still, it pays to be cautious while fishing roadside venues, as well as near swamps. Even the borders of major highways harbor venomous species. PAT FORD

An angler hefts a largemouth bass he caught while canoeing. It pays to concentrate on shorelines and drop-offs, too, if you manage to find one. The latter are usually located between 3 and 10 feet from shore, depending on the water depth and slope of the bottom. Emergent vegetation often harbors fish, as does the floating stuff—swamp cabbage, in particular. Rocky shorelines where forage can hide rank among the best places to fish—forget the lack of vegetation. STEVE KANTNER

I fished in some pretty wild places, until a rattlesnake taught me to be more careful the hard way. Now I refuse to walk where I can't see my feet, and I try to avoid areas of tangled overgrowth, along with sunny banks on cool, winter days. I never put my hands into nooks and crannies or grab bushes or tree limbs before examining them first—especially when launching or landing my canoe. Even the mowed grass that borders major highways harbors its share of rattlers and moccasins. So if you plan to walk it, wear protective boots.

While on the subject of creatures that bite, bugs in the Glades are a lot more tolerable than most folks imagine. They come out at dusk, which unfortunately is when I'm heading in to fish. Unlike in the salt marsh, where the swarms are pervasive, canals in the interior are usually fishable without annoying swarms of bugs. I minimize any nuisance ahead of time by spraying my shirt with insect repellent and stuffing it into a zipper-lock bag before placing in my car's rear window. The heat vaporizes the DEET (the active ingredient), which permeates the fabric even deeper, making it more mosquito-proof. Several manufacturers market outdoor clothing with the repellent already in the fibers.

Alligator Alley and Tamiami Trail

several major highways either traverse or border the Everglades, and they provide access for fishermen. I discuss all of them below. While the majority of anglers fish the swamps from a motorboat, you'll find ample opportunity as a walk-in angler—especially if you bring a canoe.

Alligator Alley

Arguably the most popular highway with fishermen is the portion of Interstate 75 that connects Fort Lauderdale on the Atlantic Coast with Naples on the Gulf. Appropriately nicknamed "Alligator Alley," this thoroughfare is an important conduit for

Diana Rudolph taps the Tamiami. STEVE KANTNER

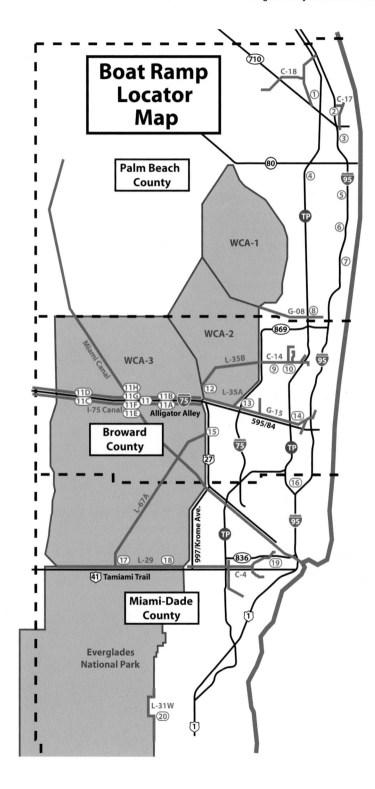

gamefish as well as vehicles, and portions are bordered on both sides by canals. What makes the Alley so special? It's the northernmost highway that crosses the Glades. Spring fishing, as you'll recall, moves from north to south.

The exits that anglers fish most often are located within the first 30 miles heading west from the Lauderdale tollbooth. Unless you're willing to pull over in 70-mile-per-hour traffic, park in a ditch, crawl through a tangle of matted undergrowth, and climb a barbed wire fence, docks and ramps at the various launch sites are the best option, especially in spring when the fish are everywhere. I've caught plenty of fish from these floating structures, including hundreds of largemouths, peacocks, and even pickerel. Several seawall parking lots stretch for hundreds of yards, so an angler can cover both banks in solitude. An even better approach is to get your hands on a canoe or bass boat and work those banks you can't reach from shore or from the ramps. As for a canoe, you'll find plenty of fish within paddling distance, if they're not already there at your feet. I don't kayak in the Everglades, due to the danger of alligators.

Parking

When exploring these canals on my own, I never gave much thought as to whether it was legal to pull off the road and access these waters, but I recently researched the issue and found nothing definitive except for two interesting opinions, one from a higher-up with the Department of Transportation (DOT) and another from a police officer, both of whom wished to remain nameless. The DOT source said, "If you pull off the road, be sure it's by at least two car-widths, and don't attempt it if the area's posted. If there's metal siding, stay behind it." The former state trooper said: "I wouldn't leave a car parked alongside Florida's Turnpike, where the state will tow it after 24 hours. The alley, however, may be a different story. If it was up to me, I'd simply let it slide."

Broward County Tollbooth to Route 29, including Conservation Area 3 and parts of the Big Cypress Swamp

If you're heading west across Alligator Alley, starting with the Broward County tollbooth (mile marker 26), you'll find ramps on both sides of the road at (or near) mile markers 31 and 39, as well as at the first of two rest facilities, located near mile marker 36. Look for the mile markers—painted reflective green—along the road at one-mile intervals. The canals are bordered by dense vegetation, so I recommend launching a canoe at a well-maintained ramp or fishing from one of the boat launches.

Shore-bound anglers (or canoeists) can fish the Miami Canal, which passes under Alligator Alley near mile marker 36, after exiting the alley at the first rest (not recreation) facility. This ancient conduit connects Lake Okeechobee with L-67, the north-south canal that runs parallel to U.S. 27 in western Broward County. Follow the signs once you make the turnoff.

If you follow the underpass under the alley, you'll find another ramp on the south alley canal—where the canal heads east.

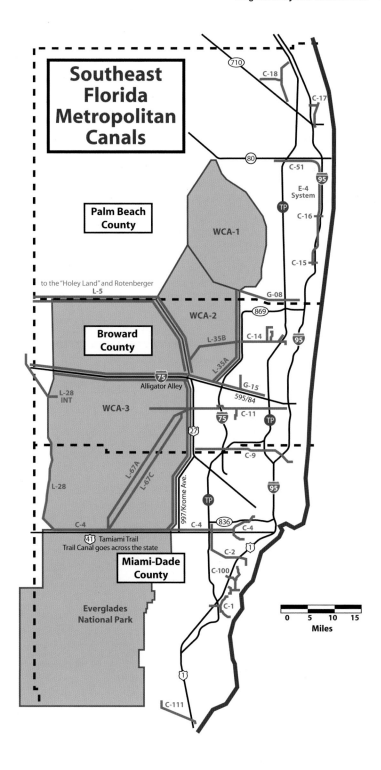

Southeast Florida Metropolitan Canals

710

C-18

C-17

80

C-51

E-4
System

95

TP

C-16

**Palm Beach
County**

WCA-1

C-15

to the "Holey Land" and Rotenberger
L-5

G-08

WCA-2

869

**Broward
County**

L-35B

C-14

95

L-35A

75
Alligator Alley

G-15

595/84

L-28
INT

WCA-3

75

C-11

TP

27

C-9

L-67A

95

L-28

L-67C

997/Krome Ave.

TP

C-4

C-4

836

C-4

41 Tamiami Trail
Trail Canal goes across the state

C-2

1

**Miami-Dade
County**

C-100

**Everglades
National Park**

C-1

| 0 | 5 | 10 | 15 |

Miles

1

C-111

If you turn around and continue heading west, you'll cross several canals, including L-28, which passes underneath Alligator Alley near mile marker 49. L-28 is located on Miccosukee Tribe land and is off-limits to the general public.

The north alley canal continues west and is less deep in the Big Cypress Preserve. This shallower portion of the canal begins at the alley's only gas station, located just off the highway at the Snake Road exit (mile marker 50). The station, which is owned by the Miccosukees, sells food and beverages. The restrooms are clean, and you'll even find showers.

While access to this section is severely limited, the L-28 Interceptor Canal (which intersects the alley at mile marker 52 and shouldn't be confused with the original L-28 that's not open to the public) offers shore-bound anglers a few miles of access. Then, both north and south the canal enters tribe land, where trespassing is prohibited.

Both the Interceptor's banks (they lie perpendicular to the alley) are reachable once you locate the turnoff, which is a pair of gravel roads that pass through a chain-link fence—not easy to see at 70 miles per hour. The gates, on either side of the road, are located above the easternmost bank—not far from the Alley bridge. Interested in crossing to the opposite bank? Take the buggy bridge a mile or two north and follow the gravel road.

A new launch ramp—complete with an exit—next to the alley bridge over the Interceptor, at mile marker 52 was under construction at the time of this writing. The ramp is on the west bank, immediately north of the bridge.

Tamiami Trail

Constructed during the late 1920s, with the intention of stimulating commerce between Tampa and Miami, the Tamiami Trail (U.S. 41) has remained a mecca for anglers—fly fishermen especially—since it first opened to traffic in 1928. It was the first road to cross the swamps.

The trail was popularized by outdoor writers who fished it during the 1950s and 60s. While the brackish portion of the Tamiami comes within a few miles of the Gulf, the first 100 miles west of Miami are freshwater habitat—similar in appearance to other Everglades canals. This part contains both bass and panfish.

Depending on water levels, fishing along the trail varies from somnolent to spectacular, with most of the action taking place during spring. In the summer, afternoon showers drown out the bite. It's a fact of life in the Glades: High water puts a damper on fishing.

The trail is an extension of Miami's SW Eighth Street, a road referred to by Miami's Hispanic population as Calle Ocho. This heavily traveled street becomes a highway and crosses Interstate 95 and Florida's turnpike before continuing west toward the Glades. Traffic on the trail is slower than on the alley, since this two-lane relic is popular with sightseers.

Unlike Alligator Alley, which parallels the Tamiami approximately 25 miles farther north, the Trail offers anglers ample bankside access. If you head west from Miami, you'll notice several makeshift boat ramps. The one Marty Arostegui and I use is located 23 miles west of the Krome Avenue intersection, just this side of the microwave tower.

Winter sunset on the Tamiami Canal. This two-lane highway has served as an artery for commerce for nearly a century. STEVE KANTNER

Anxious to ride on an airboat? Head west out of Miami on SW Eighth Street, which becomes the Tamiami Trail. The canal is in the background. MARTY AROSTEGUI

Several canals come in from the north—including L-67A, 10 miles west of the Krome Avenue Extension—and you'll find a few levees that you can reach on foot. There's a launch ramp on L-67A, just a few hundred yards from the trail. Plus, if you take the levee road even farther, you'll find a second ramp on L-67C: a canal affectionately known as the "Magic Canal" because of the trophy bass it produces.

The farther west you travel on the trail, the more bridges you'll see. While some span creeks that are essentially mud holes, others hold bass and bream. At the time

Wading for Trophy Bass

Trophy largemouth require a careful approach and proper presentation. Most trophies are spawners: females in prime condition. Here in South Florida, largemouths spawn from the full moon in November until the end of April. Waders shouldn't have any trouble spotting the beds, which are usually protected by emergent vegetation. You'll find them in 1$\frac{1}{2}$ to 3 feet of water. Look for depressions on the bottom the size of saucers. Bedding bass, however, prefer calm lagoons surrounded by grass atolls. Most spawner fishing is a matter of sight-casting with weighted plastics. However, bedding fish in the Everglades also strike on top, a proclivity that attracts a small cadre of fly fishers.

The largemouth bass is the king of our natives. Great bass fishing in the winter and spring is what draws most freshwater fishermen. Bigger bass go for a full-size meal—with the possible exception of bedding fish. For trophies on the prowl, use a magnum plug or a large plastic worm. PAT FORD

of this writing, the federal government is raising up the Trail's eastern portion in an effort to facilitate sheet flow: natural runoff headed for Florida Bay. The areas around bridges—like those near the Turner River campground, or Halfway Creek—harbor fish during late spring when the current is running.

You can usually find somewhere to park near these bridges, which, here in the interior, are typically short ones. Just be sure to pull off the road completely. Fishing from some spans is prohibited (there is no room for standing), so the best way

Approach bedding fish with caution, while determining if the lady of the house is home. Spawners, for the most part, are easy to spot when they hover close to the nest. That, however, isn't always the case, so make sure that no fish are guarding the bed before moving on. I'll toss something squirmy to bedding fish and then allow it to settle while I gauge their reaction. If two fish are present, which they frequently are, I'll catch and release the smaller, more aggressive male before working on the female.

Spawners don't behave like feeding fish. They'll pick up a lure just to carry it off the bed but pay it no mind at all if they're actively spawning. While a subtle twitch may trigger a strike, the odds are in your favor if you simply let it lie. Short plastic worms and crayfish imitations are ideal for this work.

Bedding bass respond to flies, especially those patterns that mimic a threat—for example, a prowling water snake. Keep in mind when you're dealing with bedding fish that they typically hang out in heavy cover—often inside a rim of grass—which makes it difficult to present a lure with freshwater fly gear. A 9- or 10-weight makes the job less daunting, since the heavier line pulls the fly to the surface. It's also easier to cast the gargantuan flies that big bass prefer. A heavier rod is infinitely better for handling 15- or 20-pound tippets, which you'll need to extract fish from weedy labyrinths.

Another technique that's effective for spawners, or fish that have recently spawned, is to reel a 10-inch worm along the shoreline. While you'll see very few shorelines in the marsh, I remember this trick when I come across one, especially in winter when the bass are spawning. It's also effective in canals. Here's why: In canals, bass spawn over shallow gravel patches that form when the banks cave in. Since beds in these areas are difficult to spot, especially late in the day when spawners are active, you'll need to cover lots of water. And that's where reeling big worms comes in.

You'll deal with mats of vegetation punctuated here and there by openings. Say you're casting a weighted plastic—keep reeling it across the mats and dropping it into the holes. While you allow it sink, you need only twitch it (just like when you're flippin'), while staying alert for the slightest movement. Don't be surprised if a bass explodes before you reach the hole. Drop your tip before rearing back the second your line comes tight. For this type of fishing, I prefer plug-casting gear, along with a heavy, non-stretch line—say, 30-pound PowerPro.

Bass that are not spawning move around more than most anglers realize. These movements are dictated by a variety of factors, including water level, time of year, and ambient light. It's not unusual if an area that previously held fish becomes suddenly devoid of them because the bass have moved to different cover—where you couldn't buy a strike before.

to approach them is from either side, by casting beneath them or along the shore-lines. A sign will be posted if fishing's prohibited; if not, you'll probably find a sidewalk.

The trail's crowning glory begins with the salt marsh. Whether a largemouth bass or a 30-pound tarpon, predators rely on the shadows cast by bridges, from which they attack baitfish or grass shrimp. Individual tarpon prowl while snook and bass typically suspend in schools.

Fighting a snook in the Tamiami.
STEVE KANTNER

This snook hit a Squid Fly.
STEVE KANTNER

Salt marsh snook from the Tamiami. STEVE KANTNER

My strategy for casting under these bridges while remaining dry—though I wouldn't recommend it to those who are afraid of breaking their fly rod—is to stand on the abutments that frame these bridges and launch a backhanded water haul. Aim well into the shadows, at least 20 feet back.

Route 27

U.S. 27, a major north-south route, runs along the Glade's eastern boundary. From its origin in central Miami-Dade County, this one-time two-lane death trap (cars would run off the road and end up in the canal) continues north to Lake Okeechobee and parallels the lake along its western shore. From an angler's perspective, U.S. 27 provides both walk-in and boat ramp access at Everglades/Holiday Park in southwestern Broward County, Sawgrass Recreation Park in northwestern Broward, and at a number of additional launch ramps and access points on either side of the highway.

These include the Power Poles Hole, where a series of fingers extend westward from the canal that borders U.S. 27 on the Everglades side (and that ultimately turns into the north alley canal). The power lines are visible from I-75 or I-595. The south alley canal takes a turn just east of the tollbooth and becomes part of L-67, which follows U.S. 27 into Miami-Dade County.

Black crappie or speckled perch: Spawning specks during the winter really draw the crowds.
MIKE CONNER

74

Heading south through Broward, U.S. 27 borders the Glades, before turning southeast. Then, a few miles south of the county line, you can take Krome Avenue and keep heading south, while staying within reach of the swamps. Krome eventually crosses the Tamiami, before continuing south to Homestead. There's another north-south canal, L-67C, that starts just north of the Tamiami. Parts of the Tamiami and L-67C—its sister north-south canal—offer stellar fishing. Walk-in opportunities, however, are limited except near the launch ramps.

The best fishing along L-67—associated with a series of marsh access sites—takes place at least 10 miles from the nearest ramp, which is at Holiday Park. The sites aren't accessible by anglers on foot. However, a section of L-67A in Miami-Dade County—located near the Value Jet Memorial in Miami-Dade County that commemorates one of the worst aviation disasters in U.S. history—is reachable by canoe, from the Tamiami Canal.

Heading north again in Broward County, you'll find a pair of launch ramps on U.S. 27—beneath the power poles—one and two miles north of the alley tollbooth and west of the highway. If you plan to park there, secure your valuables. The stub canals, which extend between these giant pumps, provide refuge for the fish when the state pumps water south.

Whenever water management turns on its pumps, the main canals lose clarity, because current stirs up sediment. A rule to remember about Everglades bass is that they typically hit better where there isn't much current—except below spillways or culverts that are forcing out forage or below those tiny seepages that anglers know as run-outs.

As you keep heading north on U.S. 27, you'll come to a ramp on the east side at Sawgrass Park, a recreational facility that the Feds lease out. There's a public ramp half a mile farther—again, east of the highway—where you access another canal that runs north behind a tall row of grasses. Drive past the spillway before pulling over—you'll see the signs. Bank fishing here can be spectacular, but most of the fish are small.

Continue north on U.S. 27, and you'll pass a small ramp on the east side just before the highway bends to the west—approximately 3.5 miles north of the one you just saw. There's a dirt ramp on the east side if you drive 3 miles farther; then 2 more miles and you enter Palm Beach County.

You'll also pass Johnnie's Bass Hole before reaching the county line. It's located west of the highway where a fishing camp once stood (hence the name). There's a spillway north of the turnoff. Fishing there is poor, except when the current is running full bore.

If you're heading north to the lake, there's one more option (if you don't have a boat). You get there via the L-5 levee. L-5 sits practically on the county line, at the point where the power lines split. Take the levee west—drive the gravel part slowly—and you'll find bass and bream in the roadside ditches—along with plenty of snakes.

Look to your right—half a mile west of 27—and you'll see the Harold A. Campbell Public Use Area. This area, and several marshes just like it—which are known as storm water treatment areas (STAs)—are popular with ducks and wildfowl hunters alike. To hunt them you need a special permit, randomly drawn during an annual lottery.

As you head west on the levee, you'll see water on either side, but you'll need a boat to reach the best of it—unless you keep driving west.

Culverts pass under the levees. If the water's running, and it's not too high, trophy bass suspend beneath the gars. Plus, you'll usually find mudfish. There are places to launch a small boat here.

Anglers who drive the levee to the Holey Land or Rotenberger Wildlife Management Areas pass by a series of sod farms. It was near one of these farms that I had a close encounter with, of all things, a Florida panther. Although it wasn't my first encounter, it could have been my last because the cat forgot her manners and decided to charge me. She lost sight of me as she raced up the levee, which is how I lived to write these sentences. Similar encounters are extremely rare. Two Florida Fish and Wildlife Conservation Commission (FWC) officers that I met a few weeks later assured me that the panther had taken up residence in Corbett (a wildlife area in Palm Beach County), where she was eating her fill of fat wild hogs. "Better them than me," I thought to myself.

Lake Okeechobee

Get back on U.S. 27, head north, and after 12 miles you'll arrive at Lake Okeechobee. The lake is surrounded by the Herbert Hoover Dike, an old precaution designed to prevent drowning deaths in the days before radar. During one severe storm in 1928, hurricane winds forced the lake from its banks and onto the fields—drowning more than a thousand workers.

The dike is currently under repair, which puts parts of it off-limits to would-be hikers. For walk-ins, however, there's the Pahokee Pier, where anglers can catch crappie during the winter. Missouri minnows, which can be dangled beneath a bobber, are available at the pier, located at Everglades Outpost in the town of Pahokee. Take a right at South Bay and keep heading north. Marabou crappie jigs and plastic grubs are also popular—again, fished bare or beneath a bobber.

Sawgrass, Everglades/Holiday, and Loxahatchee Recreation Areas sit on private wetlands that have been leased. Holiday Park, according to *Sun-Sentinel* outdoors writer Steve Waters, was donated by its prior owner to Broward County around 1964. The County, which has owned it since, then leased the land to the FWC, which sub-leased it to a concessionaire. Now, Broward County runs the operation.

Waters believes that Sawgrass is owned by the Broward County School Board. He's sure, however, that the Board is responsible for maintaining the boat ramp there.

Loxahatchee, he claims, belongs to the State, which leased it, in turn, to the Federal Government for fifty years

These parks offer improved access to the Conservation Areas or the Loxahatchee National Wildlife Refuge. Among the amenities you'll find at Sawgrass and Holiday are airboat rides, a launch ramp, and a general store that sells beer, sandwiches and sundries. Boat rentals are available except at Loxahatchee where it's strictly canoe rentals.

To reach Holiday, take Griffin Road west past U.S. 27, and you'll come to the Holiday Park pavilion. Griffin—a crowded thoroughfare—originates 20 miles east in Dania Beach and parallels C-11, a viable fishery. From Holiday, you can launch a boat and head north or south (L-67A) or west via the weed-choked central canal. Holiday Park is cut off from C-11 (the Griffin Road canal) by a spillway.

From Holiday, anglers in boats can reach a dozen or so marsh access sites that lead to a sunken marsh. These areas lie at least 10 miles south along L-67A—too far for a canoe and unreachable on foot.

You can head north along U.S. 27 from Sawgrass or east via a canal that opens into a marsh before intersecting L-35, west of the Sawgrass Expressway. The marsh provides action for anglers afloat, once the water drops to the appropriate level—ask someone who's been there. You can't head south due to a spillway. A dike separates Sawgrass from the Bombing Range to the south.

The Arthur R. Marshall Loxahatchee National Wildlife Preserve—the smaller of the three recreation facilities—offers visitors a mix of canals and marshes. Its address is 10216 Lee Road in Boynton Beach. Florida's Non-Native Laboratory was re-located here recently.

Big Cypress Preserve

In South Florida's western periphery—beginning just south of where State Highway 29 intersects the alley—is a narrow canal that separates this two-lane road from the Big Cypress Preserve. The canal comes in from the south. There's a spillway near the eastbound on-ramp, but it's surrounded by a chain-link fence.

Highway 29 canal. To the east of the canal lies Big Cypress Preserve, part of our National Park system that supports a diversity of wildlife, including the Florida panther. STEVE KANTNER

This tiny pastoral watercourse, punctuated by spillways and salt-intrusion dams, was once a good source of largemouths and panfish—notably Mayan cichlids. The fishing, however, has since deteriorated for reasons that remain unclear.

Flowing water makes an interesting counterpoint to the cypress swamp—especially in winter when marshes are draining. A portion of canal runs north from the alley, but it's a festering mud hole that's unfit for fishing.

Anyone who fishes along State Highway 29 during winter will find a few bass—especially below water control structures or where rivulets of current exit the swamp. I see these fish as atypical largemouths that may represent a separate subspecies.

Canoes

In this book I target areas you can fish on foot, but canoes allow you to cover more water and water that you would not want to walk in (such as the Everglades canals) without making the commitment a full-blown skiff and trailer require. Also, you won't go broke getting a canoe. I use a canoe in the spring when I fish for largemouths.

Because aluminum overheats in the subtropical sun—and since I'm unwilling to take out a mortgage to buy Kevlar—all my canoes are fiberglass models that measure between 13 and 16 feet in length. As soon as I buy one, I repaint it with a shade of Petit's Polypoxy known as "Dull Dead Grass." If you can keep from frightening the wading birds, then you won't scare as many fish. I learned that trick from Keys guide Captain John Donnell.

You don't want to buy a canoe with a rounded bottom unless you're intent on running the rapids. You don't want one that's completely flat either because the drag on the water makes the craft slower and harder to paddle. Somewhere in between is a design that pros refer to as tumblehome, which combines stability with ease of paddling.

A canoe extends your range (power optional). STEVE KANTNER

Swamp life is different in other ways, too—often less ambiguous. A sign at the end of a bridge to a homestead reads: "Here's where your rights end and mine begin." Point well taken.

Conservation Area 2B

You need a boat and motor to reach the majority of marsh spots, but a good one that's accessible on foot is Conservation Area 2B, or the Bombing Range, which lies west of Markham Park in southwest Fort Lauderdale. Area 2B is visible from

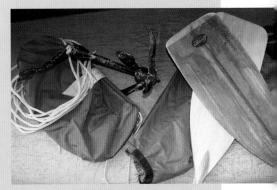

Typical canoe gear: paddle, drift sock, thwart bag, and anchor. STEVE KANTNER

Some other convenient accessories are wheels for toting canoes, as well as outrigger floats, which allow you to stand in a canoe more safely. I like Spring Creek's products (www.springcreekoutfitters.com).

If you canoe by yourself, a drift sock helps by slowing you down and acting as a rudder.

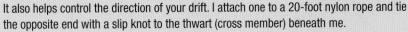

It also helps control the direction of your drift. I attach one to a 20-foot nylon rope and tie the opposite end with a slip knot to the thwart (cross member) beneath me.

I typically start by paddling upwind and then fish my way back with the breeze behind me. The drift sock helps slow me down, and I can adjust my position with strokes of the paddle. Deploy your sock or drogue in mid-canal, since the breeze usually swirls along the edges.

I always carry two paddles: a bent-shank model that acts like a supercharger on a race car engine and a standard, less expensive model. If you use a bent-shank paddle, make sure you hold it with the blade angled forward so it can grab more water. The person in the rear uses it, while whoever's up front has the basic model and helps steer the craft.

I also carry a lightweight anchor that I made by filling a one-quart milk jug with mixing concrete and shoving in a heavy eyebolt before the mixture set. It's less likely to hang on the bottom, and if it does, I'll simply untie it. Miniature foldaway anchors are also available from a variety of sources.

In addition, there's always my grass anchor—something I jury-rigged by attaching two feet of $5/8$-inch rope with a loop at one end (to loop around a thwart) and a big knot at the other that I tightly crimped to a jumper cable clip. I use the jaws of the cable clip to grab a handful of cattails. That saves me from dropping a heavy anchor, which ends up scaring the fish. You can buy the clip at an auto parts store.

I store small items in a thwart bag, a nylon pouch with a Velcro closure that I fasten to the mid-hull thwart using the same knot I use to tie my shoes. I attach it so it opens in my direction.

Brackish water marsh bass—an anomaly that differs from its sweetwater cousin. The salt marsh has bass that are fatter, darker, and a lot more muscular than their Everglades cousins—and they fight harder, too. Marsh bass feed almost exclusively on tiny minnows (making large bass baits futile) and will enter brackish water. STEVE KANTNER

The Bombing Range marsh on a winter day. A rare snail kite is perched just beyond the anhinga. STEVE KANTNER

Jack Zitt, my mentor in all things marshy, surveys the Bombing Range at dusk. STEVE KANTNER

I-75, if you look northeast of U.S. 27. After World War II, the Navy used the marsh for disposing explosive ordnance, hence the nickname Bombing Range. Their pilots got in some valuable target practice while fishermen inherited a honey hole.

Supposedly, the few trees still growing there today began their lives in the deepest bomb craters. Most of Area 2B's 1,000 acres lay swathed in aquatic vegetation. It's a clear-water habitat that's rich in forage, which encourages rapid fish growth. The substrate beneath Area 2B is so porous that it drains faster than the surrounding area—and just when the fish start biting. When that happens, the state sometimes suspends limits.

Conservation Area 2B is home to big bass, along with the occasional gar, a few panfish and mudfish, plus a pickerel or two and some blue tilapia. Whether snakeheads have arrived here is open to debate since they've already been reported from one neighboring canal: L-35.

This marsh, which is encircled by levees, fishes best when its bass population has been replenished through culverts that connect to the surrounding canals and the new fish have had time to grow—needless to say, when the water's low. If the marsh hasn't dried up for several years running, some fish no doubt will be established, instead of left to bake in the sun. This place serves as a bellwether for wading marshes in general.

You can access this marsh without getting your feet wet by crossing from Markham past a water control structure in the southwest corner. But it's easier to launch at the Markham Park ramp if you have a canoe before crossing the canal (L-35A). If you go by canoe—and you'll need a push pole—be sure to carry a flag that's attached to a 10-foot staff. It's the law of the marsh, and it's strictly enforced. You'll have to drag the canoe and your gear up and over the levee to the opposite side and pole northwest until you find the fish. I count the power poles while making the trip and attempt to line up with the fifth from the corner, looking back toward the interstate.

The bass in this wetland congregate more to the west, but move closer to the launch ramp as the water drops. Avoid float tubes and kayaks—I usually found snakes and gators near the levees.

Urban Lakes
and Canals

The South Florida landscape is dotted with waterways, including hundreds of lakes and canals. While some are located in rundown neighborhoods—where their banks are littered with garbage and debris—others support vibrant populations of gamefish that run the gamut from largemouth and sunshine bass to colorful exotics, some unique to the area. Many of these lakes and the canal systems they're part of are accessible to shore-bound anglers. PAT FORD

A few South Florida lakes are natural, as are the occasional waterways that join these bodies. Some trace their origins to ancient swamps. But the majority of suburban and urban lakes—and the canals that connect them—were dug to collect fill for construction projects. They're not spring-fed gems—or natural sinkholes—but the work of developers who needed gravel to build on or a place to dump their waste. Most of the smaller lakes were dug on purpose. They're like box-cut canals, the result of a dragline bucket.

When I was a kid, my parents warned me to avoid rock, or barrow pits. And now I understand why: swimmers can't always climb up the bank due to the angle of the drop-off. A number of children drowned in these pits, which should serve as a warning even today.

John Cimbaro, a biologist who promotes deepwater resources and the editor of *The City Fisher* (a quarterly newsletter published by the Florida Fish and Wildlife Conservation Commission for the benefit of urban anglers), assisted with this chapter. I limit my coverage of lakes to Broward, Palm Beach, and Miami-Dade Counties because that is where the important systems are located, as well as where the majority of South Floridians live and fish.

The major difference between urban and Everglades waterways is that the former aren't as affected by seasonal changes, such as dropping water levels that bottom out in May. But that's not to say that the levels don't recede as soon as the summer rains let up, just as they do in the swamps.

This bragging-size largemouth came from a drainage ditch that borders a busy, suburban strip mall. MARTY AROSTEGUI

Of interest also are several smaller lakes that aren't usually associated with major systems. Caloosa Park, in Boynton Beach, contains at least one lake that supports catfish. Heritage Park in Plantation (Broward County) supposedly does the same. Meanwhile, lakes in Topeekeegee Yugnee (TY) Park in Hollywood and Markham Park, near Interstate 595, support token populations of largemouth bass. In fact, every major park that contains a lake supports largemouths and sometimes butterfly peacocks. Peacock populations, though, were greatly reduced by the devastating freezes of 2009 and 2010. You'll also find fish in golf course water hazards if you fish early and late—but you take the risk of getting hit by a golf ball.

More information on these waterways can be found in pamphlets produced by the FWC, in conjunction with the South Florida Water Management District. Each pamphlet contains a map, along with information about boat ramps and facilities and which species you're likely to find there. All are reasonably up-to-date. You can download them at myfwc.com/fishing/freshwater/sites-forecast/s/metropolitan-miami-canals/.

Lake Mangonia

Mangonia is a large lake in eastern Palm Beach County that stretches from Palm Beach Lakes Boulevard to 45th Street and provides drinking water to West Palm Beach. Mangonia flows into nearby Clear Lake (where recreational activities are prohibited) and later into the L-8 canal. Access is limited.

According to state biologist John Cimbaro, there's a boat ramp on the western shore of the lake, along with parking and picnic tables in the town of Mangonia Park. That means there is shoreline access. To get to the ramp, take Palm Beach Lakes Boulevard east from I-95 1.5 miles to Australian Avenue and then head north 1.1 miles. The ramp's on the left. Gas-powered motors aren't allowed on the lake; only electrics.

As far as species, Cimbaro was quick to add: "When we sampled Mangonia back a few years ago, we mostly came up with bass and bluegills. However, we did shock several large sailfin catfish. Mangonia is deep; some parts are deeper than 14 feet."

But that's not where the story ends. Captain Steve Anderson grew up fishing the shores of Mangonia. Here's what he had to say:

> My buddies and I caught plenty of bass just outside the bullrushes while wading and casting with plastic worms. But our primary targets were monster [channel] catfish. We'd fish a dead bluegill on the bottom with our 10-foot bridge rods and conventional reels spooled with 80-pound-test. There wasn't much line left at the end of the cast, so sometimes those monsters would strip our reels. We caught dozens that weighed 30 pounds or more.

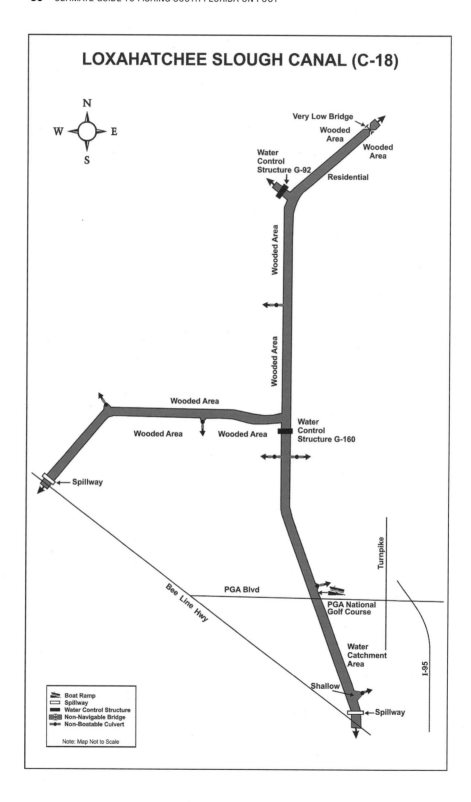

LOXAHATCHEE SLOUGH CANAL (C-18)

Martini Arostegui got quite a surprise with this trophy bluegill in a South Miami canal.
MARTY AROSTEGUI

C-18 Loxahatchee Slough Canal

Unlike most South Florida canals systems that flow through populated areas, C-18 is bordered by woods. There's a boat ramp—an FWC structure—at the northeast corner of PGA Boulevard and the main canal. It has adequate parking but no other facilities.

This canal originates in northwestern Palm Beach County—just west of the PGA National Golf Course—and runs north from the boat ramp for 5.8 miles. It forks at the 3.1-mile mark, forming a 4.5-mile-long spur that flows west before intersecting the Bee Line Highway. You'll find spillways at both the southern and western extents of this system, in the vicinity of the Bee Line.

The primary targets in this pastoral waterway are largemouth bass, bluegill, redear sunfish, spotted tilapia, and oscar. According to electro-fishing data, C-18 contains more largemouths than most southeast Florida canals, although the fish aren't necessarily larger. Here, as in other Florida waters, bass fishing improves during the winter or early and late on summer days when the water temperature drops. C-18 is close to the J.W. Corbett Wildlife Management Area, a 60,000-acre public recreational area maintained by the FWC. Corbett offers everything from hiking and hunting to a youth camp.

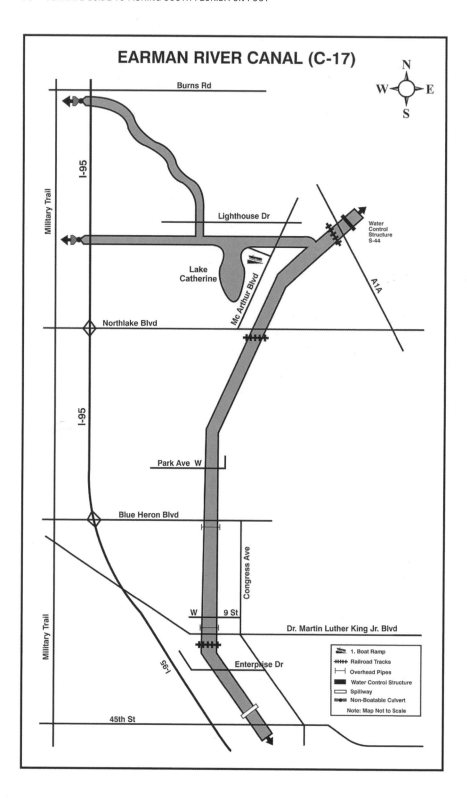

EARMAN RIVER CANAL (C-17)

C-17 Earman River Canal

This waterway runs through northeastern Palm Beach County from Clear Lake in West Palm Beach to a spillway near U.S. 1 in Palm Beach Gardens. The saltwater portion that lies below the spillway near 45th Street is famous for snook. Access, however, is limited. Above the spillway are largemouth bass (more than in most southeast Florida canals), spotted tilapia, and oscars. To find shoreline access at the canal's only boat ramp, take Northlake Boulevard east from I-95 and turn left on McArthur Boulevard.

E-4 Canal

This system, which includes naturally occurring but highly modified Lakes Osborne and Ida in eastern Palm Beach County, extends all the way from C-51 in West Palm Beach south to the Hillsboro Canal and has excellent fishing. A reliable source of

Roberta Arostegui with a sunshine hybrid she caught while fishing Lake Ida. That's guide Butch Moser holding the fish. Sunshine hybrids—not true exotics—result from crossing a striped bass female with a white bass male. The state stocks 15,000 sunshines a year in many community lakes or waters with an abundance of shad. This hatchery-supported fishery peaks from winter to late spring, when anglers free-line live threadfins or cast and troll wobbling plugs such as the floating Rapala. MARTY AROSTEGUI

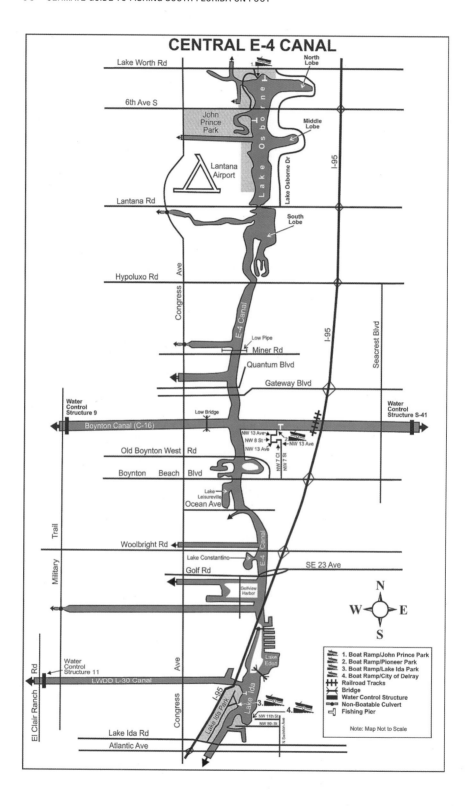

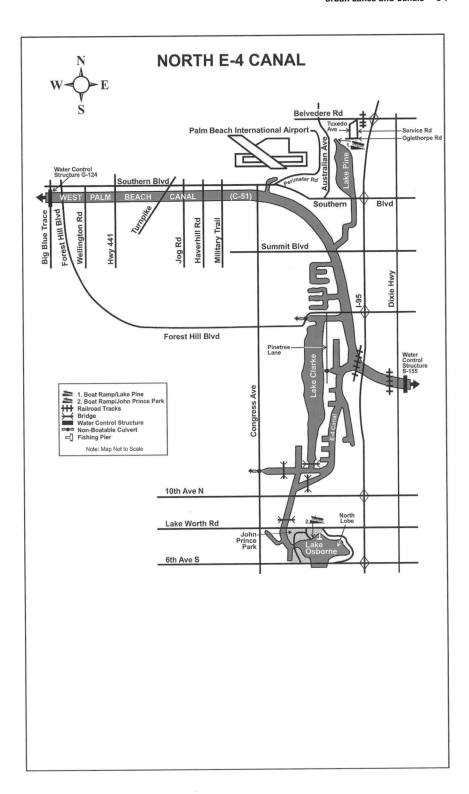

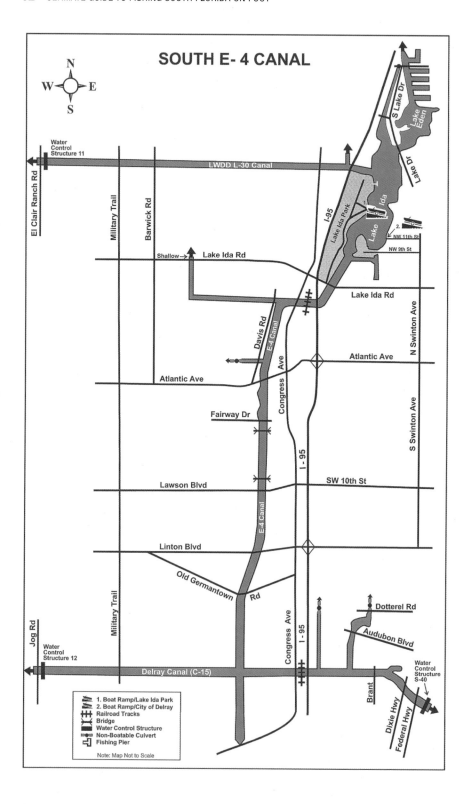

Author with a clown knifefish from Delray's Lake Ida. PAT FORD

largemouth bass, it's also the home of sunshine hybrids, along with several species of exotics. Fishing for sunshines peaks during winter, while the bona fide exotics prefer warmer weather. Four public boat ramps can be reached from Interstate 95. According to the FWC's Angler's Guide to the Central E-4 Canal, shoreline anglers will find good access on the west and south banks of Lake Ida, within the Lake Ida County Park, as well as on the east shore of Lake Ida in the City of Delray Beach. The pamphlet goes on to state that there's additional shoreline access along the east side of the canal between Lake Ida road and the I-95 overpass. You'll also find fishing piers north of the Lake Ida boat ramp, next to the boat ramp at Pioneer Park on the Boynton Canal (C-16), and east of the boat ramp on Lake Osborne. Nearly the entire shoreline of the north lobe of Lake Osborne between Lake Worth Road and 6th Avenue South is accessible to shore anglers.

C-15 Canal

This waterway divides the cities of Boca Raton and Delray Beach before flowing east to the Intracoastal Waterway. Farther west, it ties into the E-4 canal, which connects to southern Lake Ida. A major attraction downstream is the Boca Spillway, where saltwater predators gather during freshwater releases. The spillway is accessible from U.S. 1.

G-08 Hillsboro Canal

The lower half of this waterway forms the boundary between Palm Beach and Broward Counties. G-08 flows eastward from the Loxahatchee National Wildlife Refuge in southern Palm Beach County to a water control structure near Military Trail before eventually working its way to the Intracoastal Waterway. The lower 10 miles of the canal are navigable.

Several laterals enter G-08, beginning just east of Water Conservation Area 2. While the canal's western reaches are primarily pastoral, residential developments crowd the banks closer to town. G-08 hosts not only freshwater natives, including largemouth bass and several species of panfish, but also exotics, including butterfly peacock, spotted tilapia, and Mayan cichlid. Anglers who fish there may encounter snook and tarpon that worked their way through the locks.

Before that portion of C-14 lying just below the easternmost spillway was closed to the public, another of my friends caught several world-record snook there when the summer rains had started and the locks were open wide, spilling forage into the maelstrom below.

Ubiquitous Mayan cichlids hit most baits and lures, regardless of size. PAT FORD

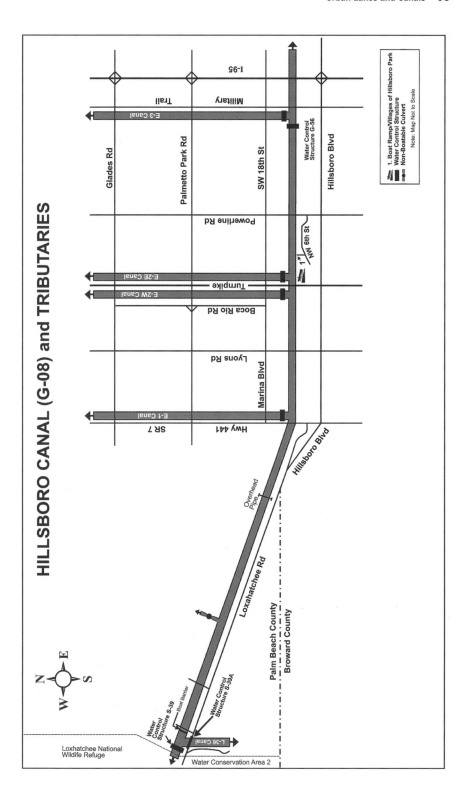

HILLSBORO CANAL (G-08) and TRIBUTARIES

C-14 Canal

C-14, also referred to as the Cypress Creek Canal or the Pompano Canal, flows east from Water Conservation Area 2—actually, from a north-south canal named L-36— to a point just upstream from Lake Santa Barbara in Pompano Beach. A major player in northwest Broward, C-14 hosts native largemouth bass, panfish, and threadfin shad, along with butterfly peacocks and bullseye snakeheads. You'll also find saltwater interlopers, including snook, tarpon, and the occasional jack or ladyfish.

The canal is divided by several spillways, the easternmost of which lies downstream from a stretch of Dixie Highway just north of NW 62nd Street, near the border of Pompano Beach and Fort Lauderdale. It's on private property behind Pine Crest School.

To get to the second and, I believe the most important, spillway, enter the Palm Aire condominium and golf course complex from Powerline Road. Take Riverside Drive and remain innocuous. While I understand that it's legal to fish the strip of Water Management land located just below the spillway, you might end up dodging golf balls.

After a thunderstorm, Palm Aire turns on. I can't count all the tarpon I've released there on flies. Plus, one of the largest largemouths I've ever seen spit my buddy's plastic worm back in his face—just as he was about to "lip" the lunker.

At C-14's westernmost terminus is the third of these spillways, one that practically borders the Everglades. Bullseye snakeheads have breached this barrier and entered the north-south canal. That canal turns into L-35 as it continues southward along the Bombing Range. Before reaching L-35, C-14 spawns several laterals,

Tiny tarpon work their way inland, usually after a major storm. PAT FORD

CYPRESS CREEK CANAL (C-14)

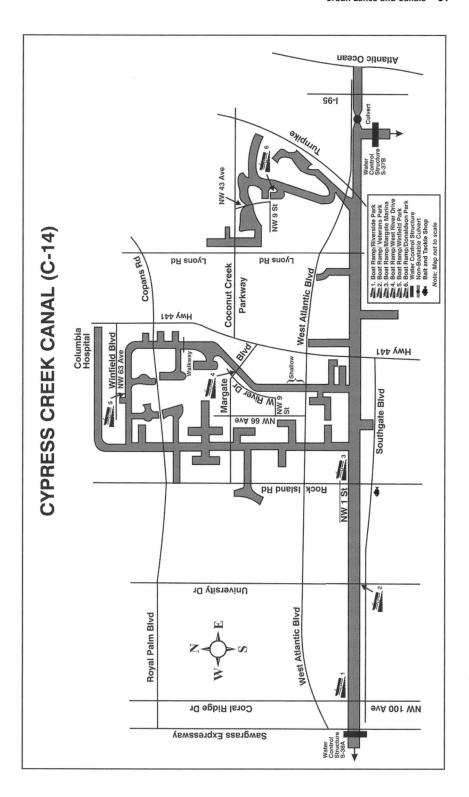

1. Boat Ramp/Riverside Park
2. Boat Ramp/ Veterans Park
3. Boat Ramp/Margate Marina
4. Boat Ramp/West River Drive
5. Boat Ramp/Winfield Park
6. Boat Ramp/Donaldson Park
   Water Control Structure
   Non-boatable Culvert
   Bait and Tackle Shop

Note: Map not to scale

Martini Arostegui, with a bass from a honey hole that my friends call the "Snake Pit" due to its excellent snakehead fishing. It's just feet from a busy street. MARTY AROSTEGUI

most which lead to the north. One place you'll find them is between Rock Island Road and U.S. 441, and there are more just west of the turnpike.

Take NW 62nd Street west and you'll discover a ditch that parallels a strip mall, a park, and several apartment complexes. It lies just west of where 62nd Street crosses Florida's turnpike. Sluices such as this one—and you'll find others if you look—are home to the infamous bullseye snakehead. Contrary to what many anglers believe, the sluices also support trophy largemouths and peacocks as well as grass carp.

This system, if you count the laterals, extends for miles. Major thoroughfares that intersect or run alongside it include West Atlantic Boulevard, Southgate Boulevard, and Copans Road in Pompano Beach. Six launch ramps provide access to boaters, while entire sections are accessible on foot. However, since parts of C-14 flow through residential neighborhoods, you'll want to be careful where you go. You don't want to end up trespassing in someone's backyard, especially since this area has a reputation as one of the most crime-ridden parts of the country.

C-13 North and South Fork of Middle River

Known as the north and south forks of Middle River, this meandering waterway splits in two just west of I-95. One arm continues eastward south of Oakland Park Boulevard, while the other does the same north of NW 19th Street (and later NE

This art deco bridge captures the spirit of the "island city." STEVE KANTNER

16th Street). The two join forces before becoming Middle River just west of the federal highway (U.S. 1) bridge. The "island" that results from this geographical intersection gives residential Wilton Manors its nickname "Island City."

On the South Fork I've caught jacks, snook, tarpon, ladyfish, mullet, and even grass carp while casting from seawalls with fly gear. Alan Zaremba reported good fishing for peacocks near a golf course that's located just west of Florida's Turnpike.

Jacks are easy to spot. They raise a ruckus when they run the seawalls in their relentless pursuit of forage. It's a spring and fall ritual that takes place on either tide, usually around the midpoint. I add two hours to the high or low at Port Everglades to figure the tide where I live just east of I-95. West of the spillway near NW 31st Avenue, tidal fluctuations are undetectable.

Juvenile tarpon also work these seawalls, mostly on hot summer nights. While they feed on baitfish in the glow of streetlamps, they remain curiously selective in this protein-poor environment that hovers precariously between salt and fresh.

When predators—say, jacks, snook, tarpon, and ladyfish—swim upstream on the incoming tide, they typically do so along the right bank (looking inland). They then reverse the process when the tide starts to ebb. This phenomenon is known as the Coriolis effect. It works just the opposite in the southern hemisphere. Ladyfish, which so far I've said little about, have no set pattern but are more common during the winter. Like all pelagics, they follow the baitfish.

If C-13 has a crowning glory, it's Cherry Creek. In my opinion, this several-thousand-foot natural tributary flowing through the City of Oakland Park is the most pristine in-town waterway in all South Florida. The wilderness section of this pastoral waterway resembles a postcard. The current is swift, which makes it unique, for the juvenile gamefish these billows nourish. I hope, by the time this book appears in print, Cherry Creek is a nature preserve. The fish are too small to catch.

Cherry Creek: A bit of wild nature in the heart of a city. STEVE KANTNER

There's another tributary flowing under the God-knows-how-many-lane super-intersection of Andrews Avenue and Oakland Park Boulevard. While the surroundings look interesting, the water is not.

C-12 Canal

This ditch flows through the city of Plantation, Florida, before continuing eastward. Parts of it parallel both Sunrise and Broward Boulevards. Like C-13, it originates from a residential canal that flows southward after crossing C-14. C-12 joins the New River just east of I-95.

This nondescript flow—part of which is a tributary of the New River—once hosted a mother lode of juvenile tarpon. But in the wake of the construction of I-95, access became a major problem. Plus, this was hardly a neighborhood to fish on foot.

New River Proper and the North New River Canal

When I was younger, both the New River proper and the North New River Canal made you think tarpon. The river itself earned quite a reputation (Herbert Hoover used to fish here), but as the fishing diminished, interest died. Then, several years back, a 150-pound tarpon was landed from someone's backyard.

Tarpon inhabit both the river and the canal now, but in reduced quantities. However, the problem with the river, from a walk-in standpoint, is that the banks are lined with homes. We used to fish near Route 84, but everything's been walled off by chain-link fences. There's a park there, too, but I haven't fished it since the 1970s—back when the land was vacant. You'll still find tarpon west of the interstate, where the river takes on an artificial character. That's also where the canal branches off and follows I-595 and, eventually, I-75.

My fondest memories of the North New River Canal are of releasing grass carp west of U.S. 441. I also caught tarpon on Marabou Muddlers, across from Markham

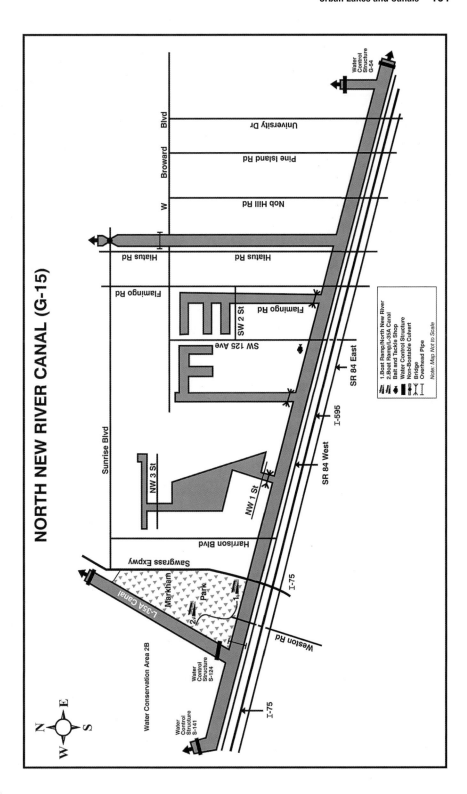

NORTH NEW RIVER CANAL (G-15)

A fly fisher supports a grass carp prior to releasing it. That's the proper way to lift heavy fish for a photo. STEVE KANTNER

Park and the Bombing Range. The canal takes a turn just below a spillway on U.S. 27 at the start of the Everglades. I've enjoyed good tarpon action there, whenever the spillway was running.

The trick to fishing this spillway is to stand on the platform that extends above the pipes. Quarter your fly down and across, and let it etch a wake near the closest bank. I've seen tarpon by the hundreds rolling in unison, but only in the presence of fast-running current. You can catch snook on a live bream or warmouth fished on the bottom.

C-11 Canal

An established favorite with roadside anglers, due to ease of access, C-11, also called the Griffin Road or South New River Canal, hosts largemouth bass, shell-crackers, peacocks, and grass carp, as well as a few outsized tarpon and, occasionally, snook. I landed a tarpon that weighed close to 70 pounds on fly gear there. I recall another slightly smaller tarpon that I beached on 4-weight gear.

That particular winter, schools of tarpon were steamrolling up and down the canal. Some fish were huge. I struggled into position by jogging and crawling—the fish wouldn't hit if they saw you coming—and launched a cast. Within a few strips, that fish was on it. Although the canal was wide, motorists on Griffin could watch the action. One pulled over and hopped on his pickup, where he proceeded to open a beer.

He kept up the pace, knocking down several more beers, while the fish grudgingly yielded across the canal. The beer drinker finally drove off in a cloud of dust—only to return with a female companion, along with a second six-pack. It took me nearly an hour to subdue that fish and to slip my fly from its lip—to a standing ovation from across the canal. Call me a show-off, but I loved the attention.

SOUTH NEW RIVER CANAL (C-11)

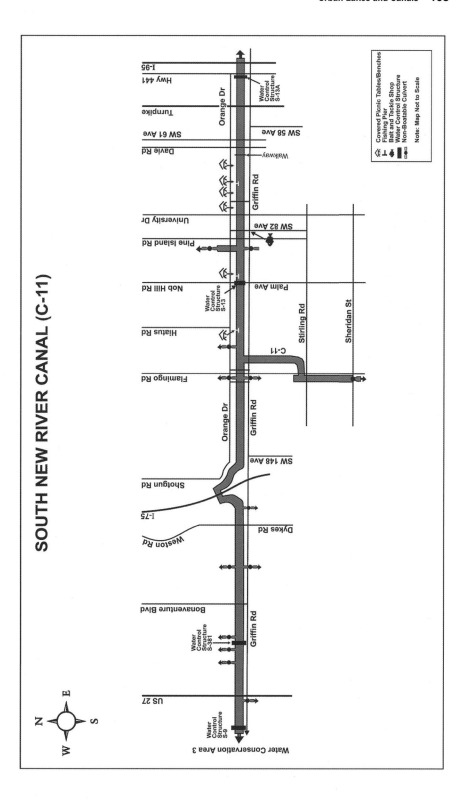

A fly fisher displays a butterfly peacock she caught in C-11 on a chartreuse Estaz Minnow.
STEVE KANTNER

C-11 was a hot spot for tarpon, and it still can be when conditions are right—meaning when the current is traveling from west to east and the water's rising. C-11 is punctuated by several spillways. In fact, whenever I go there I discover a new one. But it's the one at the intersection of Nob Hill and Griffin that, to my thinking, divides the canal in half into eastern and western portions. Each displays a different character, including different amounts of vegetation. As a result, the fishing may vary from one portion to the other. Local fixture Fred Ade fishes C-11 more than anyone else and catches tarpon at Nob Hill when the spillway's running, while I prefer to fish farther west.

C-11 is renowned for grass carp and peacocks. In fact, I once appeared on Andy Mill's TV show in an episode titled "Grass Carp King." The fame didn't go to my head. Rather, traffic personnel curtailed my future prospects by installing more guardrails and limiting the places where I can pull off and cast. Despite these limitations, there's plenty of green space—across the canal on the Orange Drive side. Walk-in fishing is hardly a growing industry because the local authorities do little to encourage it. You'll find several "piers" jutting into C-11 at various points along its length.

C-9 Snake Creek Canal

Situated on the border between Broward and Miami-Dade Counties, Snake Creek flows from a water control structure at U.S. 27 that borders Conservation Area 3 to a salinity control structure east of Dixie Highway. Included in this drainage is a series of laterals, along with several lakes. Snake Creek hosts butterfly peacocks, largemouth bass, jaguar guapote, Mayan cichlid, spotted tilapia, and oscars. You'll also find tarpon and snook that probably swam upstream through the spillway when it was open.

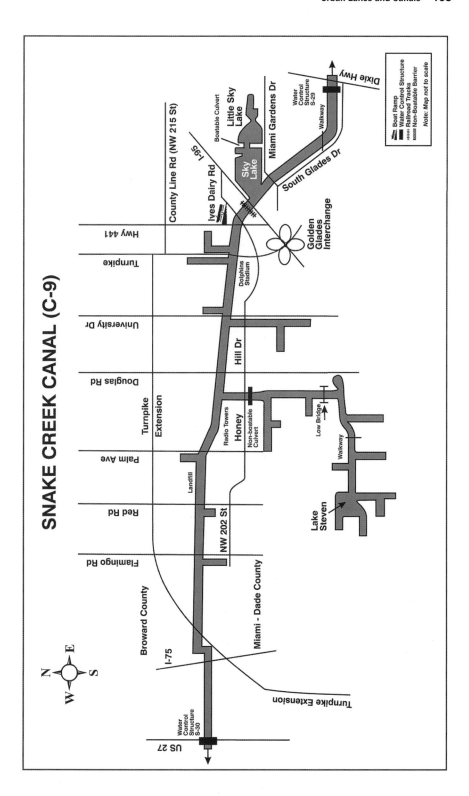

SNAKE CREEK CANAL (C-9)

C-4 Miami Airport Lakes and the Tamiami Trail

C-4 is an extensive system. From Conservation Area 3 and the Everglades, it flows all the way to the water control structure near I-95 and the Dolphin Expressway, which is just upstream from Biscayne Bay. C-4 includes the so-called airport lakes, portions of which are visible just north from Miami International Airport. These lakes are surrounded by condos and apartments.

Plenty of open water is available to boaters. Yet along much of C-4, shoreline access is severely limited. This is Miami, and almost every parcel of land has been developed.

C-4 contains peacocks, as well as largemouth bass, oscars, spotted tilapia, and jaguar guapote. You'll also find saltwater interlopers, including snook, tarpon, and jacks. Though the official state pamphlet shows the system ending less than 30 miles from where it begins, I've fished it all the way across the state (including the salt marsh portion, where a dredge dug it).

I've fished this system with Alan Zaremba by boat. However, we discovered peacocks in several locations that could have easily been reached from shore. From the launch ramp off NW 7th Street in Miami, C-4 eventually parallels SW 8th Street—a major east-west artery that continues past Krome Avenue and into the Glades.

There's a place along the way where the power poles change from creosoted wood to concrete pillars and a culvert comes in from the north. I've caught peacocks and tarpon during the middle of the day there. Steve Waters, the outdoors columnist for the South Florida *Sun-Sentinel*, was with me once when I landed a tarpon that hit a marabou slider.

Mike Conner, creator of the Glades Minnow fly, used to fish it at night before moving to Stuart. The snook, he claimed, were eager to hit, but gators were waiting and so was the traffic. He had to park on the median strip.

Back when I first started guiding, I'd taken a client all the way to the salt marsh. The weather turned cold, so despite our efforts, he had nothing to show for a day's worth of casting. I suggested we drive home on the Tamiami Trail and fish this culvert in a last-ditch effort. I figured that the deepwater canals that were closer to Miami would be warmer than those in the salt marsh. It turned out I was right, and my client landed a 40-pound tarpon—his first, I recall, on fly.

Also attached to this system are the Snapper Creek Canal (see page 111), which runs along the Turnpike Extension, and the Coral Gables Canal, which parallels the Palmetto Expressway before heading east past Ludlum Road.

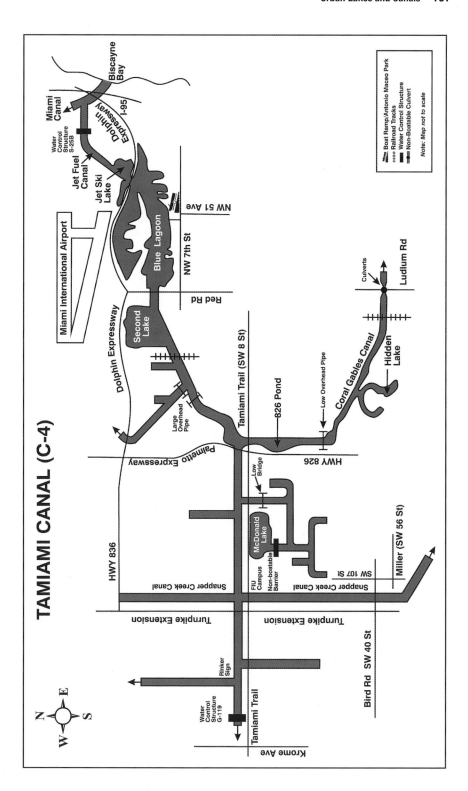

TAMIAMI CANAL (C-4)

C-1 Black Creek

State biologist Paul Shafland released the first butterfly peacock into Black Creek in 1984. The canals in this region are all part of the Biscayne aquifer, in which the water temperature remains fairly constant, even during periods of prolonged cold. C-1 is a trifurcated canal that splits in two near the turnpike extension before spawning C-1N near Eureka Drive. The main artery eventually empties into Biscayne Bay via a water control structure near Black Point.

In addition to butterfly peacocks, C-1 hosts largemouth bass, oscars, spotted tilapia, and brilliantly colored Midas cichlids. Snook are also present throughout the canal, making it possible to catch what the FWC calls a "canal trifecta" (peacock, largemouth, and snook). C-1 hooks up with C-100 via a water control device. The species of fish that inhabit these two systems are essentially the same.

Boulders have been piled up to block the canoe ramp on Black Creek; a fence prevents access to another launch facility. And the crackdown isn't limited to that canal system.

I recently spoke with Alan Zaremba, and he commented on the loss of access on Black Creek. As he put it: "I called the FWC about it, and they claimed they weren't aware of it. But I can't help wondering if they and water management aren't trying to keep us from fishing there."

Midas cichlids prefer clear-water canals, like C-4 or C-100 in South Miami. PAT FORD

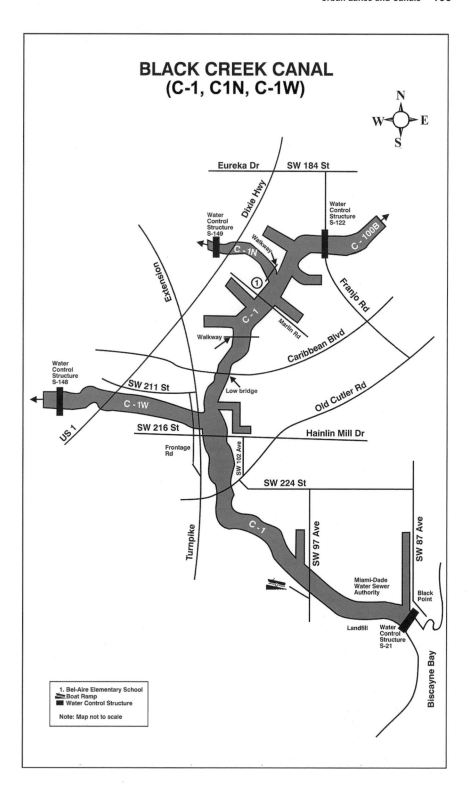

BLACK CREEK CANAL
(C-1, C1N, C-1W)

1. Bel-Aire Elementary School
Boat Ramp
Water Control Structure

Note: Map not to scale

Marty Arostegui displays a world-record peacock he caught in the suburbs. When fishing from the banks, cover as much water as possible unless you can see your targets. MARTY AROSTEGUI

C-2 Snapper Creek

Located in the central Miami-Dade County cities of Sweetwater and Kendall, Snapper Creek flows eastward from the intersection of the turnpike extension and Palmetto Expressway to the S-22 salinity control structure on SW 57th Avenue. Snapper Creek is a popular peacock hang-out.

Cutler Drain Canal

This canal in southeastern Miami-Dade County has always been one of my favorites. It's where I caught my first peacock. The canal was blasted through coral rock, which explains why the water remains so clear. While this canal is difficult to fish from shore since the majority flows through residential neighborhoods, you'll find several sections that are accessible on foot.

The 13.5 miles of navigable water in this system are officially divided into four canals: C-100, C-100A, C-100B, and C-100C. C-100B is separated from the Black Creek Canal (C-1) by a water control structure near Eureka Drive. C-100 empties into Biscayne Bay not far from Old Cutler Road.

There's a makeshift ramp across from the Falls—the shopping center where Flip Pallot once owned a fly shop. You can fish the banks to the east on foot if you park near the ramp or drive downstream to the park. In addition to peacocks, you'll see Midas cichlids standing out like sore thumbs in the clear water. Add largemouth bass, spotted tilapia, oscars, and Theraps hybrids to this hodgepodge, and you'll

SNAPPER CREEK CANAL (C-2)

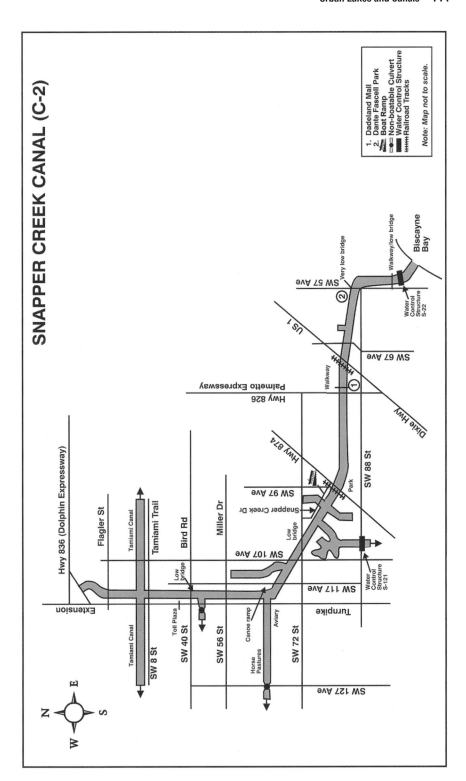

Legend:
1. Dadeland Mall
2. Dante Fascell Park
- Boat Ramp
- Non-boatable Culvert
- Water Control Structure
- Railroad Tracks

Note: Map not to scale.

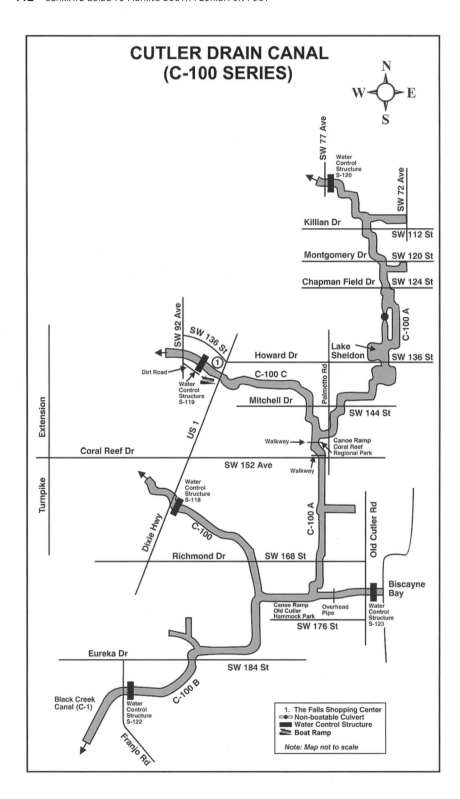

find plenty to look at in this natural aquarium. A few snook also enter this system when the spillway is open.

Two shallow lakes, Round and Sheldon, are also part of this system. While the canals average 12 feet in depth, the lakes are slightly shallower. I've seen tarpon roll there.

I caught my first peacock near a school and park complex, which I found by exiting the turnpike extension and heading east on Coral Reef Drive (SW 152nd Street). I had gotten the tip from Andy Novak, who owns LMR Tackle in Fort Lauderdale. According to the map, there's a walkway, along with a launch ramp for canoes.

C-111 Aerojet Canal

This, the southernmost of all water management waterways, looks more like the Bahamas than a freshwater canal. That's because the limestone substrate acts as a filter. C-111 is home to butterfly peacocks, as well as largemouth bass, spotted tilapia, Mayan cichlids, and oscars. The location of this canal between Homestead and the Keys explains the warmer water temperatures during cold snaps. This helps ensure the peacocks' survival.

Alan Zaremba, who has probably spent more time fishing water management canals than anyone, spoke of all the shoreline access that's available on C-111N, a part of the C-111 system that lies farther east. The Z-Man said something similar about L-31, a canal running north and south, that's located just east of Everglades National Park.

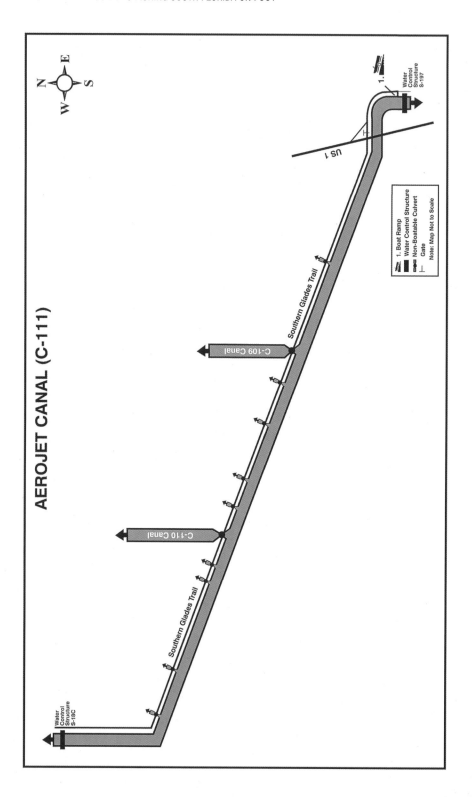

The Great In-Between

PART III

An Ecosystem in Flux

Along the east coast, the in-between zone extends from just west of the Intra-coastal Waterway (ICW) to the eastern border of the original Everglades. Lengthwise, it stretches from the C-111 canal, which passes under the U.S. 1 bridge halfway between Homestead and the Florida Keys, to the Lake Osborne canal system, near downtown West Palm Beach. Heading south from Naples, it extends several miles inland from the Gulf's inner bays to where the water remains perpetually fresh.

While mainstream gamefish from both fresh- and saltwater—including snook, tarpon, redfish, and largemouth bass—enter brackish haunts, this peripheral zone also hosts exotics and migrants, which heighten the excitement of any outing. Brackish water's salinity levels vacillate between 15 and 17 parts per thousand, depending on rainfall and the stage of the tide. The farther you get from the salt, the fresher the water generally becomes.

Canal tarpon hit flies, lures, and live bait, too. PAT FORD

Retired thoracic surgeon Carl Gill dissects Halfway Creek while searching for tarpon and snook. The occasional redfish is another possibility. STEVE KANTNER

It's this salinity gradient that encourages gamefish to leave the salt—in pursuit of forage, more suitable water temperatures, or relief from parasites—while freshwater species do just the opposite. Anglers can take advantage of these mini migrations of snook and tarpon, along with jacks, ladyfish, and a few token redfish. Meanwhile, canals below spillways may be glutted with predators following freshwater releases. Summer downpours, especially those associated with tropical activity, cause the spillways to open and canals to foam. Examples can be found from Miami to Fort Pierce—thanks to Lake Okeechobee—where venues like the South Fork of the St. Lucie River, both Lake Worth and Boca spillways (with public access), and the Palm Aire complex (which is water management land, but still accessible, although it's surrounded by condominiums) provide predictable and frequently stellar fishing. The upstream side of these concrete barriers around the spillways also hosts saltwater migrants, along with the usual suspects. In Gulf Coast creeks, sheepshead and drum join the upstream migrants. A number of brackish honey holes, mentioned in the salt marsh chapter, can be accessed from shore.

Take the salt marsh proper, which lies roughly in between the Everglades and Gulf of Mexico: It's a several-mile-wide strip of tangled vegetation that loosely parallels the inner bays of the Gulf and consists of prairies, strands, and hammocks. Strands and hammocks are areas of higher elevation that support different flora and fauna than the surrounding areas. Much of the salt marsh lies within striking distance of several major roads. Salt marsh per se doesn't appear on the map, yet terms like "marsh" and "mangrove" dot southwest Florida's periphery. The area I'm referring to stretches from just east of Ochopee to the easternmost environs of the city of Naples and extends several miles inland from the Gulf of Mexico's inner bays.

The most famous salt marsh artery, U.S. 41 or the Tamiami Trail, earned a reputation by the 1940s as fly-fishing pioneers drove over from Miami. In addition to the trail, highways 29 and 91 bisect this brackish hinterland, providing anglers with additional access—by simply stepping from their cars.

Marshes tend to look alike. However, brackish habitat can easily be distinguished by the presence of salt-tolerant plants, specifically mangroves and buttonwoods. The most visible among them is the ubiquitous red mangrove, a tree that perches atop multiple prop roots. You won't see mangroves—which are unrelated—along Alligator Alley or east of Ochopee along U.S. 41. That's because both these venues are freshwater habitat, although the salt keeps working its way farther inland.

The late Wayne Janes, whose family was among the first to farm this region, explained that when he first worked his gas station—on the corner of the Trail and Route 29—there weren't any mangroves (or snook) across from his store. But that was before saltwater intrusion. His family helped install the removable salt dams that you see in the canal along Route 29.

Brackish habitats ebb and flood with the tide. When saltwater creeks become fresh too quickly, their inhabitants adjust by moving downstream. It may only require a minor adjustment, but if the change is too rapid, the fish disappear until conditions return to normal. Or they're displaced by other species. Certain clues announce such drastic changes. Any lasting alteration in the streamside vegetation suggests such a shift. This occasionally happens when the flow is rerouted—say, after a major storm. Extremely clear water is another indication that salinity has dropped.

There's more to the in-between than just mangroves and salt creeks. Consider the hundreds (or even thousands) of seawalled canals that flow through South Florida's sleepy suburbs, some which I covered in the freshwater chapters, while others represent true saltwater habitat. Key species that I target in the surburbs are peacock bass, as well as the vegetarian trio of grass carp, tilapia, and mullet, and snakeheads and cichlids. Other oddities (including giant armored catfish) have occasionally strayed from their normal habitat (say, in Biscayne Bay or downstream from spillways). Most are dead or dying by the time anyone finds them. Meanwhile, a freshwater species—the Florida spotted gar—is known to survive below running spillways as long as the water remains partially fresh. These rascals are an indigenous species that proliferated exponentially with the digging of Everglades canals.

A Gallery of Exotics

Though you'll find exotics in the state's interior—including butterfly peacocks, oscars, Mayan cichlids, Chinese grass carp, snakeheads, blue tilapia, and armored catfish—the in-between hosts the majority of South Florida's exotics.

Florida's earliest recorded exotic is an aquarium favorite known as the oscar. Taxonomically speaking, it's a cichlid. Other interlopers joined the oscar, including both Mayan and Midas cichlids; blue, spotted, and Mozambique tilapias; the jaguar guapote and Theraps hybrid (both cichlids); along with the bullseye snakehead, the Chinese grass carp (or Amur), and ultimately, the peacock bass—not a true bass but

A South American native, the oscar—which inhabits fresh water as well as slightly brackish habitats—showed up in South Florida during the 1950s. Oscars in the wild (no pun intended) prefer banks with access to marshy areas. PAT FORD

a more aggressive cichlid, one that was introduced to control populations of other nonnative species.

Exotics compete with native species to the detriment and possible extinction of the latter. In a state where recreational fishing generates over a billion dollars annually in revenue, that's unacceptable. According to Paul Shafland, former director of Florida's Non-Native Laboratory in Boca Raton, "The danger of unauthorized stockings cannot be overemphasized." Not only do the interlopers compete for food with the natives; they vie for breeding space.

The list of exotics continues to grow, and the farther south you go, the more species you find. These aquarium castoffs proliferate as a result of owners sparing their pets a toilet flush or fishermen establishing their own private fisheries within the confines of their favorite waterways. Either way, laws prohibit such releases. Unauthorized stockings are a violation of state and federal statutes, and the possession of a live prohibited species—such as the bullseye snakehead—constitutes a felony. More significant is the risk to the aquatic environment, which is why the penalties are so harsh.

Peacock Bass

Peacock habitat was thought to be well-defined until a recent discovery just east of Naples left the biologists guessing. Currently, the northernmost range of the peacock is Delray Beach in Lake Ida.

The peacock stocking, which was Shafland's project, took place after years of painstaking research. After raising three generations of disease-free offspring from

The butterfly peacock is a beautiful, hard-fighting gamefish—easily accessible and eager to strike. The larger speckled variety was also stocked in South Florida during the mid-1980s.
MARTY AROSTEGUI

eggs he'd acquired from South America, he released peacock fry into two Miami-Dade County canals back in 1984. Both C-1 and C-100 are within the Biscayne aquifer, where groundwater seepage moderates temperature extremes. Shafland theorized that the stockings, which took place between 1984 and 1987, would control burgeoning populations of oscars and spotted tilapia that infested local waterways. He was right on both counts. Nowadays, peacocks help keep "bad" exotics at bay, while providing sport to a cadre of followers.

Despite rumors to the contrary, Shafland and his staff insist that only the butterflies (and none of the speckled variety) survived for more than a few years after the initial stocking. Peacock guide Alan Zaremba—perhaps the world's leading peacock guide, based on the numbers of days he fishes and his overall totals—claims that one of his clients caught the last two specimens back to back.

I doubt if anyone fishes more days for peacocks, either here in South Florida or in South America, than Alan. He has a four-rod rule that I tease him about, but there's nothing funny about the results he achieves. The idea behind the rule is that, at any given time, peacocks may exhibit one of four behavior patterns: spawning or getting ready to, guarding their nests or young, roaming the canal edges and actively feeding, or schooling during the fall and winter.

A peacock's mood can change in a moment. Fish that hit on top in the early morning may become lethargic as the day wears on. In contrast, the butterfly peacock is one of few freshwater fish that's known to become more active as the water warms.

The peacocks created a premier fishery. Thanks to Shafland and his crew, it's now possible to enjoy fishing in Miami-Dade, Broward and Palm Beach Counties that was only available—with a few exceptions—in remote jungle settings in South America. Peacocks, grass carp, bullseye snakeheads, and clown knifefish are unable to tolerate even moderate salinity, while oscars and Mayan cichlids will stray into brackish water. Another factor that limits the spread of butterfly peacocks is the species' intolerance to cold.

Chinese Grass Carp

Grass carp (also known as Amur) were introduced in hopes of ridding our waterways of pernicious vegetation, yet they eat more than just weeds: They also eat the berries that fall from streamside ficus trees. I imitate these berries with spun deer hair imitations that I recently upgraded in favor of 10 mm cork balls. Dr. Otto Lanz, my fellow "researcher," taught me to paint them with Avon Cherries Jubilee nail polish. I mount mine on #4 Gamakatsu SC15s.

It's illegal to possess a grass carp without a permit. They have delicate gills, so the best way to control them while you attempt to unhook them is by gripping the base of their tails after pinning their lower jaw with a Boga-Grip. Keep them cradled and in the water.

Other tricks that ensure a grass carp's survival include avoiding them in extremely hot weather and gently releasing them into subaqueous vegetation after pointing their noses into whatever current may be present. Keep them out of the sunlight by leaning over them while removing the hook. As for specific hot spots,

Grass carp management is currently under the aegis of Florida's Department of Agriculture. The fishery is strictly catch-and-release. PAT FORD

they change with the landscape, as ficus trees are planted or succumb to the weather or development. The prime times for berry-dropping are September and mid-April—right around tax time.

Snakeheads

According to Dr. Walter Courtenay, formerly of the U.S. Geological Survey, snake-heads could wipe out an entire native species by crawling from one body of water to another—on their outstretched pectoral fins—before eating their new neighbors out of house and home. Dubbed "Dr. Frankenfish" after sounding the alarm following the discovery of a separate species (the northern snakehead) in a Crofton, Maryland, pond, Courtney is now heralded as a world expert.

Shafland and his crew don't share Courtenay's fears—at least not to the same extent. They carefully monitored South Florida's snakeheads and attempted to determine if snakeheads, which are bound to spread, were harming populations of native fish. Based on their studies, bullseye snakeheads, although unwanted, don't appear to be having a measurable impact on native species, including popular sport fish. I know several locations where snakeheads are common but so are bluegills and largemouth bass—individuals representing classes of several different years.

These fish no doubt feed aggressively at times—especially if left undisturbed. But the majority of *Channa mirulius* are confined to the C-14 system, which meanders through residential northern Broward County—hardly the place to strut on

The bullseye snakehead is a recent arrival, originally native to Southeast Asia, that may have spread from aquarium releases. Bob Newland, the assistant manager at a local park, landed Florida's first snakehead in the 1990s. STEVE KANTNER

Marty Arostegui's favorite snakehead lures are the black Zoom Horny Toad or a saltwater popper. What else will a snakehead hit? You name it. While they're partial to poppers, say a size 1/0, sliders have also proven effective, as have Deceivers or dark-colored Glades Minnows. Marty's strategy is to cover maximum water while casting ahead as he walks the banks. He makes vigorous surface retrieves, the way you would for a hungry largemouth. MARTY AROSTEGUI

your fins: an accusation made by snakehead critics. If anything's for sure about this species, it's that they prefer squalid canals, especially near culverts.

Once a snakehead sees you walking the bank, the odds of it hitting are greatly reduced. However, early in the spring, when snakeheads guard their young, you can tip the odds in your favor. Try sinking your streamer beneath a big snakehead before nudging the fish gently by lifting your rod tip. Sometimes the fish will rise up and gulp your fly before settling back to the bottom. They're different than peacocks, which—believe it or not—are more aggressive. When you come tight to a snakehead, hang on for dear life.

Bullseye snakeheads are now routinely reported from L-35, the canal that borders the eastern edge of Conservation Area 2B (the Bombing Range). Report all suspicious-looking catches to the State's Non-Native Laboratory Visit www.myfwc.com. Kill the fish whenever possible, and be sure to take plenty of photos.

Jaguar guapote. PAT FORD

Theraps hybrid. MARTY AROSTEGUI

A pair of pacus parts a school of bluegills. PAT FORD

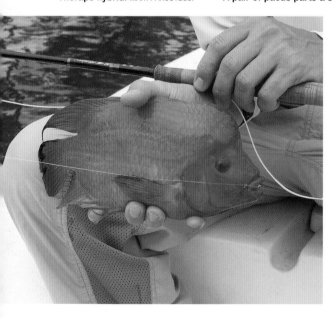

Midas cichlids prefer clear-water canals, like the C-1 and C-100 in South Miami. For a Midas, you'll want something small and chartreuse, say a 1-inch curly-tail grub tail mounted on quarter-ounce jighead, or a Midasizer that incorporates a rotating spinner blade. While oscars and Mayans slam good-size lures, an eighth-ounce chartreuse jighead tipped with a 1 1/2-inch chartreuse curly tail ranks among the deadliest weapons in the war against exotics. Peacocks love 'em too.

STEVE KANTNER

Knifefish, imported from Southeast Asia, grow to 10 pounds or more. They are in several Palm Beach County lakes, (primarily Lake Ida), and I've heard of them caught in Alligator Alley and a canal in Miami. Anglers catch these novelties from residential docks on swimming baits (Rat-L-Traps, Rapalas, streamers), but most are landed on slow-trolled live threadfins. MARTY AROSTEGUI

Spotted tilapia prefer vegetarian fare except when they're breeding in spring, when they'll occasionally strike a streamer—especially a Woolly Bugger—in an attempt to drive the interloper off. The larger blue tilapia's 2-foot-wide nest resembles a bomb crater. KELLY GESTRING/FLORIDA FISH AND WILDLIFE CONSERVATION COMMISSION NON-NATIVE LAB

Saltwater lionfish—an Indo-Pacific species—are well-established here, and there's even a lionfish derby intended to control them. If you ever come across one, proceed with caution. Do not attempt to grab it! The lionfish's fin rays are like tiny hypodermics that inject a powerful toxin. Like most small reef fish, they hit natural baits. Some consider them gourmet fare, but there's the haunting specter of ciguatera—a potentially deadly toxin that may be present in the meat. Indo-Pacific lionfish pack a wallop, as well as a threat to Florida's marine environment.
MARTY AROSTEGUI

Bedding Peacocks

The best opportunities you have for catching and releasing a trophy fish will be when peacocks are spawning. The first thing I do is look for a bed, which shows up as a light spot within a few feet of shore, usually in less than three feet of water. Beds are not much larger than a cigarette pack, and they may be on top of a culvert, or anywhere else where the bottom's flat. Next you have to attract their attention—a piece of cake, if you know the routine.

First, let's do it with fly gear: Butterfly peacock strike flies, as well as lures, so when blind-casting, I use a Clouser Minnow variant that I tie on a size 2 or 1/0 Mutu Light hook with light barbell eyes—or a weighted Flashabugger. Cast either to wherever you think a peacock is hiding before allowing it to sink and starting a fast-strip retrieve.

Cast your fly and allow it to sink directly on top of the bed. Spawners, for the most part, aren't frightened by humans, but it helps to avoid any lateral movement. Be sure that your fly actually touches bottom, while resisting the temptation to twitch or strip it.

At this point, a peacock usually picks it up and carries it off the bed. If you strike too soon, the hook pulls out or you'll end up snagging the fish. Keep repeating the process, while removing the fly before either fish is able to grab it. They'll then become agitated—their colors light up like a Christmas tree.

Wait a few seconds before casting again; this time, aim your fly just beyond the bed and allow it sink all the way to the bottom. (Try not to hang up.) Now point your rod tip directly at

Frank Cassidy displays a fly-caught peacock he tricked in a south Broward canal.
STEVE KANTNER

Captain Butch Constable, who guides out of Jupiter, shows off a peacock he caught in a southwest Broward canal. He used spinning tackle to land this fish and several others.
STEVE KANTNER

your fly and begin a fast-strip retrieve. That's when most peacocks hook themselves. In the unlikely event that a fish loses interest, switch to a different color (say, from purple or black to a chartreuse Bugger) and start over again.

As with all spawner fishing, I use at least a 15-pound-test tippet and quickly release any fish that I land. Sometimes I yield to my ego and shoot a few photos for me or my clients.

Hardware fishermen who target spawners should also cast beyond any beds they encounter—with a deep-running wobbler (say, a Rapala Shad Rap or Rat-L-Trap)—before reeling their lures directly across them. Bedding fish hit the first time around—and with trebles, the hooks stay put.

Back when I guided, I used pointless plugs to locate spawners in deep or high water for fly fishers who'd then pursue them with flies. I'd cut off the hook points before starting to prospect. Removing the hooks altogether changes the action. Peacocks, remember, prefer clear water, where it's easy to spot a fraud.

Falling Water

The lackluster landscape of the salt marsh, although varied with prairies and hammocks of old-growth cypress, is interlaced with creeks and canals. Along with stay-at-home species, the creeks and canals host migrants that move back and forth between the Glades and Gulf. These infusions take place according a schedule that takes its cue from the weather.

Regardless of the season, you'll find flora in the salt marsh that are unique to tropical or subtropical habitats, including the red mangrove, a shrublike tree that sits atop multiple prop roots that extend like the legs of a chair or table. You'll find several lookalikes, including the buttonwood and black mangrove—the latter sprouts pneumatophores that push up through beaches. The rarer white mangrove prefers life upstream, where salinity levels are typically lower. Red mangroves are salt tolerant and play a role in shaping the land—by collecting debris while they filter the water—which contributes to shorelines and establishes islands. Plus, their roots host a variety of tiny organisms—some which are part of the food chain—while providing a nursery for juvenile predators, including redfish and snook.

Red mangrove roots serve as cover for those same predators once they come of age. While the rubbery foliage shades both baitfish and predators, it's the plant's filtering and nurturing effects that make it special in this part of the world.

In addition to the more obvious signs of feeding fish such as pops and swirls, look for feeding fish in bends or creek mouths and recesses beneath the region's bridges. While a hot spot may produce for a time, it will most likely change with the water level. Creek mouths, however, tend to remain reliable, as do drop-offs and

Red mangroves perch atop multiple prop roots, while tarpon roll in the foreground in Halfway Creek. STEVE KANTNER

Wading birds gather near run-outs in backcountry canals once the water starts dropping.
STEVE KANTNER

holes in the litter-strewn bottom. Current speed determines where predators take up positions, but it's not as important as whether the water's dropping.

Forage gets trapped when the marshes drain, and a falling tide helps speed up the process. Where better to hunt than over a baited field? The birds know this all too well. I follow their lead, while looking for portals where marshes empty and gamefish gather. These run-outs set the stage for most roadside activity. While locations may change or flows become blocked—usually following a spate of bad weather—sooner or later, every marsh has to drain.

Run-outs serve as conduits for the natural chum line that salt marsh predators instinctively respond to. Snook are the most common predator, and their muted coloration and flattened bottoms are ideally suited for backcountry creeks. Plus, the position of their eyes on the top of their heads helps them focus on forage that swims near the surface—which the majority does in these shallow haunts. While snook are considered to be nocturnal, they often don't act that way here in the backcountry. I've seen them quit at dusk, only to start feeding again a few hours later.

You can't beat feeding birds for finding the action—during daylight, of course. Bingeing birds suggest that there are fish around, and the more birds you see (and the more aggressive they are), the more snook or tarpon will usually be in the fray.

The further the water falls, the more forage it moves, forage which ends up in creeks and canals.

Think about the decreased salinity that results from freshwater runoff. It must change the taste of the water. Northerly winds blow water out of the bays, thereby increasing the 12-foot gradient leading to Lake Okeechobee. The tide and blue skies also have an effect.

Exact scenarios depend on ambient water levels, which take their cue from seasonal runoff. So if the water's already low, like it is in March and on into May, then it doesn't take all those factors together to create a similar effect—a tip worth remembering if you're into springtime fishing. Water levels in these canals fluctuate independently, and the salt marsh is dynamic. Beginners seek hard and fast answers, the kind they get from a GPS, while seasoned anglers analyze subjective data. Tidal movements, water levels, and water clarity (predators prefer it slightly off-color) are worth more to them than the best of charts.

Try keeping a log to improve your fishing. One way to do this is to purchase a tide guide and make notes at the end of each day's fishing. Learn to look at tides as mini roller coasters, and plan to arrive at the start of the fall—preferably after the highest high when it first starts to drop. Entries should include the wind—both its direction and strength—along with the temperature. Water temperature is even better. Then correlate your notes with peak times and results. Here's what you'll find: The more moving water—if it's not too cold—the more action you'll see on a given day. Generally, on cold days, salt marsh gamefish prefer midday temperatures; otherwise, dawn and dusk.

A ditch I refer to as the River Styx takes its name from the hellish bugs and brambles as well as sticks on the ground that make fly-fishing difficult. Still, I know one spot in particular where a creek enters Styx, which is reliable for hooking a few tarpon—if you catch the right tide. But the problem I have with some tiny creeks is that they're so well-hidden that even I can't find them. It pays to line-up these hot spots with a visible reference, like a telephone pole or tree.
STEVE KANTNER

Captain Dave Hunt with snook. STEVE KANTNER

Tide Lag

Salt marsh tides run several hours behind the closest reporting station, which is typically near the coast. A mangrove lake that's two miles inland has tides that may run as much as two to three hours later than what you see in the tide guide. While it helps if you have a map—specifically, of the area between Naples and Chokoloskee—an aerial photograph is even better. Both will make it easier to follow the fish. But first, you must decide where to start.

Slight variations in the distance inland equate to major differences when it comes to the tide. I call this "drag time." Tides primarily depend on longitude as well as how far you are from the coast. Take a several-mile stretch of Route 92 between Marco Island and the trail. The current in those canals that border 92 may be running one way close to their outlets and the opposite close to the trail.

Don't forget that you're trying to intercept gamefish that are riding the tide in or out. Look for natural signposts—that may have nothing to do with feeding birds—where fish may be visible. Schooling species, such as jacks and tarpon, pass by landmarks—say, a dead tree or creek mouth—at a specific phase of the tide. Identify the latter and you outflank the fish.

Winds

Weather is critical. A southerly or westerly breeze is the kiss of death along Florida's southwest coast, but after it changes to a more northerly direction, the fishing ramps up in the neighborhood of five or six hours later—depending on how hard it blows. That's exactly what happens when a cold front passes through.

Because of the lay of the land, westerly or southerly winds are a bane, since they pile up water against the coast, which holds back the tide. Wind from other quadrants allows it to drop, and a breeze from the north helps it fall.

Water Temperature

Water temperature is important, too—snook and tarpon prefer temperatures in the mid 70s to lower 80s Fahrenheit. Tarpon don't mind if it gets slightly warmer, but if it drops too much below their comfort zone—as it does during prolonged winter cold snaps—both species will vacate the backcountry. The smart place to look for them when readings drop below 70 degrees is wherever you can find either cloudy water, a dark-colored bottom, or, if you know such a place, a warmwater spring.

I'm reminded of a mud flat where the locals ride their swamp buggies that's easy to see from land. The good news is that, along with the fun they're having, all that sloshing stirs up mud, which enters the main canal on a falling tide, where it's carried by the current to a nearby bridge. That makes the water there several degrees warmer than the surrounding area. On a chilly day, that can make a difference. I recall fishing that bridge with a magazine editor, and that temperature difference rescued our trip. We released several tarpon on a day that I would have written off.

Seeing Red

The majority of reds taken from roadside canals are caught during fall and winter—when the water temperature drops below 70 degrees. Since the snook and tarpon have left, there's not much else to compete for your fly, except a few largemouth bass and a gar or two—and gars become dormant when the temperature drops.

Most roadside reds are immature and called puppy drum. However, Fort Lauderdale gastroenterologist Dr. Alin Botoman, released a copper-flanked beauty one memorable day that weighed in the neighborhood of 15 pounds. A remora was attached to its side at the time—proving how recently it had left the salt. It hit a marabou slider on Botoman's last cast.

There were other times I encountered nice-size redfish—once while casting a surface plug at a well-known clearing on the easternmost canal on Route 29. You can recognize it from the moldering boat wreck.

Red drum (or redfish) are usually found over grass flats in Florida Bay and the Indian River Lagoon. However, some move inshore to spend the fall in the salt marsh. PAT FORD

Fresh water turns to brackish whenever runoff and salt water mix. If the meeting is facilitated by a northerly wind, there's an immediate uptick in feeding activity—especially if moving current's involved. Look for snook and tarpon where bass held court, along with other euryhaline species (those capable of acclimating to either salt or fresh water). Donna Ross hooked this tarpon in a backcountry creek where current was flowing. STEVE KANTNER

Cold Weather

Cold weather presents a major challenge, depending on not only how cold it gets, but also how long it persists. I remember when water temperatures remained in the mid-50s for ten days in a row along the trail and in the surrounding creeks. That's cold enough to support Atlantic salmon, and it killed plenty of mangroves. Someone told me it actually snowed in Chokoloskee.

It takes two incoming tides (each bringing with it warmer water) after the wind finally shifts into the east and the temperature moderates to return the snook to their former haunts. Heat-loving tarpon may stay away for months until the water warms. Just the reverse takes place during summer: That's when the canals get too warm for snook.

High Water

Say you're crossing the trail in July or August and the water's a foot or two higher than it was last winter. That's when you face the problem of runoff: the result of recent rains. Rain, you could say, is a lot like medicine, where a little is good but too much is a killer. When summer deluges persist and the canals overflow, you can barely recognize that stretch of highway I mentioned earlier.

I doubt you'll see swirls from your air-conditioned car, but any that you find can be blamed on mullet or the ubiquitous gars that inhabit this hinterland. So how can you combat these high-water blues, when the water practically laps at your tires? Try putting it to work by finding where it's moving—say, at a culvert or spillway.

With the air temperature hovering in the mid to high 90s, it's hardly the season for snook. They typically vanish by the end of May and gather in the passes prior to spawning. Yet juvenile tarpon that weigh from 2 to 30 pounds look forward to summer. While they prefer to frolic at dawn and dusk, they give their best performance after afternoon showers. They poke their dorsals through the film when the surface slicks off. Other herring-like fish do much the same, including bonefish, ladyfish, and American shad.

When the canals you've been fishing are deeper, juvenile tarpon have a greater sense of security. Unquestionably, the best place to find them from July through October is where you find current, say, on the downstream side of a culvert or in canals that enter the trail from the north.

You may have to embark on an extended search, but it'll pay off in the end. Options on the trail include the first ten or so bridges west of the Faka Union Canal, along with the deeper channels east of Highway 29. The meadow portion of Halfway Creek—immediately downstream from the trail—has tarpon of all sizes, but only at certain times. Deep water there provides the ideal refuge until they decide to forage along the mangroves.

When I'm tired, which I don't mind admitting at the end of the day, I'll park next to a bridge and roll down my window (watch out for bugs!), while I wait for the traffic to pass. Then, when the noise abates, I'll listen for pops. This is when the tarpon are just warming up. If the sun is still high, I'll peruse the mangroves and under the bridge in the shadows. I pick a spot with falling water, which beyond Faka Union, typically heads west. West of Highway 92, the direction reverses.

Also worth checking are the earthen crossovers that lie north of the trail on Birdon Road, east of Highway 29. Remember, current is what triggers the action, along with sufficient depth. While Wooten's on the trail gets its share of traffic since tarpon in the big lake are plainly visible, usually they suffer from lockjaw. And they also tend to stay out of range, so don't be tempted to trespass—not with so many other viable options.

Another area worth checking lies across from Wooten's, where tarpon can be seen rolling on the falling tide. Ladyfish are there, too, when the tide is falling. Then there's a waterway just east of Big Cypress Preserve Headquarters, where park volunteers leave their RVs. Because of vehicle restrictions that apply to the rest of us, you have to fish it on foot.

You'll find numerous waterways in this bug-bitten world that you'd assume are salty—if for no other reason than they're clear and green. That assumption, however, may be incorrect, since the local limestone substrate acts as a filter. So the brown stuff you see in a particular canal may actually be saltier than the green stuff next door. Predators, you'll recall, prefer off-color water.

I've run across fish and wondered how they got there. Maybe a flood or through an underground channel? I'm reminded of tarns that lie south of the trail, near the tiny hamlet of Ochopee, or behind the Indian Village on U.S. 41, just east of the Faka Union Canal, where the tarpon appear to be totally landlocked.

I once considered the weir at Faka Union a waste of time despite its fishy appearance. Now one of my cohorts tells me otherwise. He remains the only dissenting opinion. Continuing development seldom helps the fishing.

If you check out these far-flung locations, you may have some walking to do. But don't be surprised if you turn up silver. Don't rely on egrets in summer,

Tom Greene with a snook from the Boynton spillway. No one, to my knowledge, has devoted more time to spillways—or enjoyed greater success in the process—than Greene, who has lived in South Florida since the late 1950s. TOM RYAN, COURTESY OF TOM GREENE

though: most fly the coop until the temperature drops.

In this part of South Florida, something is always brewing. When fall rolls around and summer storms abate, the cycle begins anew. Then it's back to snook and the egrets and, later, redfish and salt marsh largemouths.

Equipment Essentials

There's no clear demarcation between brackish and fresh water in the marshes of southwest Florida. The water's salty one day, and the next it's fresh, which can mean a change in forage (albeit a minor one) that requires anglers to change both their lures and their tactics.

Tackle for this region is similar to what works in the swamps. The exception is at the larger spillways, where anglers dangle gizzard shad and other large live baits on saltwater winch gear and plug casters toss lures on rods that are stouter and sometimes longer than they'd use in the swamps. In addition to the same incidentals that I recommend for fresh water, you also need to have both a saltwater and freshwater license when you fish brackish venues like the Tamiami Trail canal. But you won't need a Snook Stamp unless you intend to keep one.

Whether in the salt marsh or suburbs, I cast small Rapalas, weightless shads, and marabou jigs. So give me that 7-foot spin rod once more, matched with 6- to 10-pound mono. When I'm plug casting, however—say, with a Shad Rap, Rat-L-Trap, weighted grub, or a jig—I like a saltwater outfit spooled with 20-pound mono. I've hooked monster tarpon in suburban canals and landed some big ones. Plus, at those larger spillways, I have a second option: Fishing a shad or mullet on 60-pound bridge gear and getting my butt kicked.

Fly anglers should fish heavier leaders, such as a 7-footer with a 16- or 20-pound tippet (Mason hard mono). With larger flies—meaning an occasional 1/0—I may even add a 30- or 40-pound shocker. Two-section leaders work fine.

Tips on Technique

Most salt marsh predators feed by cornering their victims and attacking them either from behind or beneath. Certain natural barriers facilitate their efforts—say, fallen cattails or shoreline rocks. Look at those mollies and guppies that exit the marshes through conduits that I refer to as run-outs: Snook, especially, know their whereabouts. Whenever a snook finds forage in sufficient numbers—typically during the outgoing tide—it races through the school, inhaling a mouthful, which often results in that telltale pop. It sets off a chain reaction, and within minutes the water is alive with explosions.

Snook appear less active while the water's rising, although all these rules are made to be broken. I repeat that disclaimer throughout this book, but here's the truth: I know a place where snook frequently feed on the incoming tide. Snook are known as "ambush predators," meaning they ambush forage that's swept down past their lairs. Tarpon, on the other hand, attack from beneath. Plus, they'll feed on either tide—as long as it's moving.

While forage doesn't literally sweep downstream, the current does help align it, which makes predation a lot easier. The baitfish concentrate on staying in formation while predators approach from blind spots. Predators attack tightly packed schools so they don't have to expend more calories chasing food than they'll gain in return. They're a lot less likely to pursue individual baitfish—unless those individuals will provide sufficient calories. The tiny flies and lures that work in the salt marsh, imitating the prey there, are like needles in a haystack, unless they are presented with a wake.

Snook prefer shorelines that are lined with cattails (including those surrounded by copses of mangroves), while tarpon focus on deeper redoubts—often in the vicinity of red mangrove prop roots. The recesses deep beneath the roots rank high on the list of tarpon stopovers. The same can be said for snook and creek mouths. Meanwhile, both join forces beneath the area's bridges.

The major goal of any pelagic predator is to capture maximal forage with minimal output. They do it with an exaggerated sucking motion, which makes a popping sound. With snook, that fatal gulp is preceded by a rush; with tarpon, by a spiral up from beneath.

The best all-around way to fish the majority of flies—especially flies like the Rivet—is to cast down and across before retrieving them partway back with a series of twitches. Aim as close as possible to the opposite shoreline, where most strikes occur. And don't waste time trying to fish out each cast. I strip the fly by pronating my wrist while pulling with my line hand—a short, sharp pull. That adds plenty of extra snap, which makes all the difference with finicky fish, and it doubles the number of fish you'll hook. Since salt marsh predators typically hug the far bank (possibly in response to traffic noise), the beginning of the retrieve is the most important. Don't waste your time on unproductive water. Just pick up your fly and cast it again.

With certain patterns, it's a different story. Buggers fish best in slow currents, or where there is no current. Fish hit poppers and sliders right away.

Lures need to be lightweight and small: soft plastics, marabou jigs, the tiniest of crankbaits, and the occasional straight-running topwater plug (such as a Zara

Puppy). With an unweighted shad, keep the twitches light, while if you're using a crappie jig—say, in a deep canal—barely jiggle your rod tip while steadily reeling.

While this fishing may not be sight-casting in the classic sense, you're casting to targets: breaking fish or suspected lies. After your fly hits the water, allow it to sink just long enough to get beneath the surface. Then, immediately begin a short strip (or with a lure, short-twitch retrieve) while letting the current enhance the action.

Quarter downstream and you add to this effect, which may be the only way to get a strike. But every so often, especially with tarpon, fish will grab the fly when it hits the water. Whenever that happens, strip-strike.

Spillways

I include spillways in this chapter, but they are also important in fresh water. South Floridians are deluged by rainwater—5 feet of it every year—which refuels our aquifer or eventually drains out to sea. South Florida Water Management District canals facilitate the process, and following heavy rains, those below spillways run awash with foam. While runoff still floods a few low-lying areas, spillways (and levees) help limit the damage and direct the excess to where it's less of a problem. An added bonus is the stellar fishing that takes place below these concrete buttresses.

The author hefts snook from a suburban spillway that he caught and released on a two-handed fly rod. Like the tarpon he was targeting, it hit a Marabou Muddler. STEVE KANTNER

Look at any water management map (like the one on page 67), and you'll see flood control structures spread throughout Martin, Palm Beach and Miami-Dade Counties. The water managers open them according to an irregular schedule that, to critics, defies conventional logic. Usually, however, it's because of thunderstorms, which, by the time June rolls around, are regular occurrences. The threat or presence of a tropical system assures that every floodgate in South Florida will be open. Nobody wants a flood. Since spillway fishing takes place in the worst possible weather, come prepared. For real time information regarding spillway openings, check the water management's live data web page at www.sfwmd.gov/portal/page/portal/levelthree/live%20data.

It's a familiar drill: The downpour abates, leaving suburbia soaked while off in the distance, machinery rumbles. The floodgates open at a neighborhood spillway where hyacinth-choked runoff starts surging downstream. Itinerant gamefish travel upstream to meet an easy meal amid the glut of baitfish that wash through with the torrent.

It's a fast, furious, and fatal encounter that seldom lasts more than a few hours. Things really start hopping both in the suburbs and nearer the coast, where spillway releases can be truly spectacular. Snook are the major target where spillways drain into the salt. It's not uncommon, however, for them to travel 20 miles inland, as measured from the nearest inlet—practically to the border of the Everglades.

Tarpon are another attraction beyond just snook. The percentage of silver kings (or kinglets) increases the farther inland you go. These sometimes-landlocked predators feed on tiny shad and *Gambusia* baitfish. Meanwhile, much-larger live baits and magnum lures provide most of the action near the ICW, where gargantuan forage is typically the rule.

This is the upstream side of a water management spillway near the ICW. STEVE KANTNER

Torrent gushes from the St. Lucie spillway, part of a move to lower Lake Okeechobee. Prolonged releases have a disastrous effect on South Florida's ecology.
MIKE CONNER

Fishermen generally ignore smaller spillways tucked back in the suburbs. Yet jacks and ladyfish join a few snook and tarpon there, along with true freshwater residents, such as largemouths and peacocks. Peacocks, especially, require freshwater habitat, where there's no measurable salinity. You'll only find them where fresh meets fresh.

While a few Florida gars get swept into the salt where they appear to acclimate without much trouble, certain freshwater species, such as bullseye snakeheads, avoid spillway releases—probably because of the raging currents.

Picture a number of minor spillways—ones that divide freshwater habitats. Then imagine the upstream side, where forage has been accumulating for the past several weeks. This grist for the mill includes cichlids and sunfish, both threadfin and gizzard shad, mollies and *Gambusia*, and possibly freshwater mullet. Tiny scuds and grass shrimp, if they're present at all, won't be noticed by predators downstream.

Now picture that to the west, a thunderstorm is brewing. This forage is washed downstream whenever the current in the canal overwhelms it. The storm, with its lightning, continues to build and work its way toward the coast.

The skies darken and the wind starts to howl before the subtropical deluge brings life to a standstill. While the downpour continues with no letup in sight, the water keeps rising upstream from the spillway. Then screws start to turn and the gates lift and the current swirls. Threadfin shad, a freshwater species that gathers in schools, are caught unaware and pass through the floodgates in this seething cauldron. The shad become disorientated as they're swept through the structure, although they try to regroup as they're forced downstream. The current overwhelms them when they attempt to fight it, so they turn upstream to regain their balance.

The shad are unable to buck the raging torrent, so they back downstream while trying to make it to either shore. By now, they've gathered in tight-knit schools. Unfortunately, the schools stay on top, where predators—in this case, tarpon—can track their movements. This ultimately leads to the shads' demise, as the tarpon start crashing through them. Any that survive end up hugging the banks.

A host of predators began swimming upstream at the onset of the freshwater release, their movements triggered by the taste of the water, variations in the current, or a change in the depth—who really knows?

There's a waterway near my home—a featureless monster that resembles those sluices that you frequently see in Hollywood chase scenes. Three separate spillways divide it. When the water rises, floodgates open, shunting the runoff toward the Intracoastal. I hear the huge screws turning from inside my car before I hear the groaning metal and gurgling water. Then the runoff starts streaming beneath the floodgates, and in a minute or two, the canal's awash with what could only be described as foam. This is a dangerous situation, so avoid entering the water either upstream or downstream of spillways. The state posts warnings and strings barrels across canals in the interest of protecting the lives of boaters. Here, we landlubbers have the advantage.

A hundred yards farther downstream, the surface becomes wrinkled. If you're willing to wait, the tarpon will come. It starts with a single fin. Then dozens of tarpon start rolling at once. While it's difficult to tell from where you're standing, they're feeding on threadfins that are trapped in the current. If you're already rigged, you can cover those rises, but your efforts will be in vain unless you move upstream and start quartering your casts at a downstream angle.

Fishing Spillways

I believe the major reason minor spillways get ignored is that so few anglers have any luck fishing them. Those who come armed with live bait have better luck: shad or shiners that they hook through the nostrils and dangle downstream. Or flies.

The secret to spillways lies in knowing where to fish them, which depends on where the fish are feeding. The big mistake that most spillway fishermen make is standing directly across from their quarry—if it's rolling tarpon—wherever there's a fast moving current. No lure looks natural from this approach.

Pretend for a moment that you're a hungry tarpon, holding in the midst of the maelstrom. You've probably discovered a hydraulic cushion—a place where the current isn't so strong. It may be behind or ahead of a rock, or possibly a piling or ledge.

By now you're eyeing the wrinkled surface in hopes of spotting a meal. And lo and behold, those shad reappear—concerned more about current than with critters like you. So you head for the surface and assault them. Then, after you've gulped that initial mouthful, you turn on your tail and continue the attack. Your efforts keep carrying you farther downstream, where other tarpon are joining the carnage. Shad that survive this relentless gauntlet line up along the bank, where snook and jacks finish the job.

Learn to look at the current the way a predator does. It's impossible, for example, to dash across it while all the time fighting the flow. That's why baitfish heading for the nearest bank drop downstream in their escape attempts. So in order for your offering to resemble the naturals, it must also make that downstream swing—unless you're dragging it near the bottom where the current's reduced.

My favorite tackle for fishing smaller spillways is fly gear and often a two-handed rod, which makes swinging flies and line-mending easier. As for patterns, I prefer Marabou Muddlers—I know that I'm fishing them properly when I can see a wake. The long rod helps keep the line straight. To avoid a belly, simply lift up some fly line and give it a flip upstream. Or do just the opposite, if the current requires. A mend is just a sideways roll cast that you use to reposition your line after casting. Keep your line as straight as you can to help control the fly. While dry-fly fishers aim for drag-free drifts, you want your streamer to slice the flow—while showing it to the fish tail first. When you hook a tarpon or snook on a Spey rod, it's like holding onto a fire hose.

Say I arrive at the spillway while the tarpon are rolling, armed with my trusty telephone pole. All I have to do is put it together since I've already rigged it with a fly and leader. So I twist the sections and make sure that they line up before securing the ferrules with electrical tape. Then I strip some fly line onto a grassy patch before making the first of my false casts.

I don't haul, just let line slip, since the rod does all the work. I keep one hand positioned at the top of the fore-grip, with the other on the butt below the reel seat. Once I've extended approximately 80 feet of double-taper, I allow it to fall at a downstream angle. (An ordinary weight-forward line works fine, too.)

I can barely see my Marabou Muddler or the V-wake it makes while it etches the current—which immediately pulls at my line. But before a belly can form, I lift my rod tip and flick part of the line upstream.

My fly pauses momentarily. Then at the start of the swing, a tarpon grabs it and takes off running. I've been following the fly with my rod, so the hook essentially sets itself. From there it's a battle royale as the 30-pounder vaults toward the clouds.

While hardware doesn't perform as well as flies here, it does an adequate job. Live bait is always effective—if you know how to fish it—but it's better left to those larger spillways, where heavy tackle and big snook hold court. If I had to pick a handful of lures, they'd be small floating Rapalas and weighted grubs—ones with a

The Lake Worth spillway, near downtown Lake Worth, has a reputation for monster snook.
STEVE KANTNER

swimming tail. Marabou crappie jigs are also effective, but tarpon invariably straighten the hook.

Work crankbaits as you would a fly—by casting downstream at a 45-degree angle before allowing them to swing across the current until they come to rest near the bank below you. Then reel them back. Never reel against the current (see Lures for the In-Between). I'll get to bigger baits and heftier lures when I talk about saltier venues. Meanwhile, here in the boonies, think small.

Remember, quartering downstream is only a valid technique if the current's raging. In pockets of calm, or in protected lies, cast to feeding fish like you normally do. Meanwhile, when you think about color keep in mind how black shows up against the riffled surface.

You'll find spillways scattered throughout South Florida—from the Everglades all the way to suburban developments and on toward the Intracoastal. Some—such as Johnnie's Bass Hole on the west side of U.S. 27, just south of the Palm Beach-Broward line—fish better than others. It's seldom open, but it's a magnet for bass whenever it's running. You'll also encounter a few pickerel there, along with the usual gars and mudfish. Another Glades hotspot that's seldom running, S-128, sits at the intersection of I-75 and U.S. 27. When this one's going, look for tarpon and snook.

In Broward County alone, there are dozens of spillways and culverts. Lock structures interrupt east-to-west management canals like C-14 in Pompano Beach, the New River Canal, and C-11 near Dania. There's also Lox Road. You'll find tiny spillways on residential canals, along with hundreds of culverts and lesser outflows. Look for fish wherever a small canal joins a water management conduit. If you don't see visible feeding, then take a few casts.

You'll find culverts galore in the Holey Land—a watery expanse with limited, walk-in access—except when the water's low. Look for ones where the water's running. Instead of reserving their best for the long-rod crowd, these culverts produce for bass casters armed with a jig and a pig (a jig tipped with a pork chunk or a soft plastic trailer). In the Glades, the quarries are trophy largemouths. You'll find natural drainages, too.

Anglers who travel on salt marsh roads pass several natural drainages where the techniques used in suburban settings are equally effective. You may end up downsizing, but waking flies across any appreciable current is your best bet. All you need is running water to find tarpon, snook, and sometimes ladyfish. During the rainy season, that's an easy order.

Mega-Spillways That Border the Salt

So far I've written more about minor spillways than about ones that empty directly into the salt—real salt, that is, like the Intracoastal. Included among the latter are the Boca Raton and Boynton spillways, the well-known Lake Worth Spillway—located between West Palm Beach and Lake Worth on C-51—and the one that controls releases on the South Fork of the St. Lucie River in Stuart. Both the quarry and bait are larger in mega-spillways, as evinced by a history of epic catches.

Snook are the primary targets here. However, monster jacks and outsize bluefish make occasional appearances. Go to Palm Beach County, where tarpon are scarcer,

Barriers like these barrels keep boaters from getting too close to spillways. STEVE KANTNER

and everyone targets snook. The baits used by anglers below mega-spillways—typically live gizzard shad, live mullet, and huge lures—remain consistent throughout this coverage area.

In Miami-Dade County, try the spillways in South Miami or Coral Gables where a water management canal flows into another canal or into the salt. Examples include C-1 and C-100. None that I've heard of produce consistent fishing, though.

In Broward County, try 441 and Orange Drive; the NE corner of the intersection of U.S. 27 and I-75; culverts or spillways along Griffin Road (C-11) in southwest Broward; C-14 in Pompano Beach; or Lox Road near Parkland.

In Palm Beach County, follow U.S. 1 north from Glades Road to find the spillway on C-15 that borders both Boca Raton and Delray Beach. Take 10th Avenue north to U.S. 1 and then turn right for the Lake Worth spillway. You need a boat to fish the Boynton spillway on C-16 unless you have a friend who lives downstream. Over the past several decades, the Lake Worth spillway has earned a reputation for trophy snook. There, in the midst of a bustling city, snook still hold forth when the floodgates open. Now rebuilt, this area boasts two sets of catwalks and a cleaning table for dressing your fish. Here, bluefish join snook in gorging on forage.

Other spillways worth considering include the Southwest Fork of the Loxahatchee River just off I-95 at Indiantown Road in Palm Beach County and the St. Lucie Dam and Lock, which joins the South Fork of the St. Lucie River with C-44, the Okeechobee Waterway in Martin County. These spillways fish best when the water turns salty (20 ppm).

The Salt

PART IV

The Sea and Its Environs

We South Floridians live with the salt—whether the Atlantic Ocean, our jetty-lined inlets, or an extensive network of seawall waterways. Our beaches extend for miles, providing limitless access for shore-bound anglers who are willing to walk and explore. When one spot is not producing, another probably is.

In the northern hemisphere, beaches located on the western periphery of large bodies of water, such as the Atlantic Ocean, experience more wind and wave action than their eastern counterparts. Take the surf at Hobe Sound or Palm Beach, com-

The surf at daybreak offers renewed promise. You can also fish the surf at dusk and during the hours when the tide is at its highest. PAT FORD

An easterly wind tends to kick up the surf, and any current it creates usually brings with it new fish, while the extra aeration perks-up their appetites. Rocky Fikki (in the foreground) and Terry Luneke fish the beach regularly by casting huge plugs. STEVE KANTNER

pared to Sanibel, Captiva Islands, or Florida Bay. The former are high-energy shorelines with windswept beaches, persistent chop, and the threat of groundswells.

Wave action, along with the wind and current, serves as the catalyst for shoreline fishing. The same holds true in the Gulf, where a more tranquil scenario sets a different pace. Wind and waves typically work in concert to push volumes of water along the coastline in what hydrologists refer to as the "longshore current."

Fish—at least, migratory ones—go with the flow, while stopping occasionally to feed. Along Florida's Gold and Treasure Coasts, thanks to prevailing southeasterlies and the proximity of the Gulf Stream, the current typically flows north. Occasionally, however, it turns around when the breeze shifts to a northerly direction. A burly Nor'easter—in fact, any north wind—can transform the surf from a millpond to a maelstrom in a matter of hours.

When cold fronts pass during winter, groundswells form in the Gulf Stream before crashing ashore on unprotected beaches anywhere north of Hillsboro Inlet (Pompano Beach), where the Great Bahama Bank is no longer able to block them. The accompanying surge brings fish with it, often quickly and in incredible numbers. In spring, rising temperatures trigger renewed activity, but inshore waters may take weeks to recover. Once the breeze settles into the east, migratory species begin retreating northward to wherever they'll spend the summer. Their subsequent replacement by schools of baitfish signals the start of summertime fishing. This sequence defines the near-shore realm: an area referred to as the littoral zone. Here,

Pier fishing runs the gamut from somnolent to spectacular. We've all heard it said: 10 percent of all anglers catch 90 percent of the fish. So assume that the 10-percenters are doing something right, learn what it is, and follow their example. I've concocted elaborate strategies to increase my chances—from establishing secret codes to practicing my "game face"—both work well in the surf and on piers. PAT FORD

Schools of baitfish head north as the water warms. PAT FORD

along with the sand and the seaweed, you'll find structures that appeal to anglers as well as fish. These include fishing piers, inlets, and the jetties that frame them—along with portions of the Intracoastal Waterway that are easily accessible, including a number of bridges.

The Inshore Calendar

Say it's New Year's Day and you're here on vacation. What types of fishing can you get to on foot? You've already read that the swamps are draining. However, hit-a-cast bass fishing is still unlikely until sometime in March. But on the piers and in the surf, some anglers pursue run fish, such as Spanish mackerel and pompano, while others target larger ones—like kingfish and cobia, or sharks in the surf. The wind will probably be chilly and the waves may seem daunting for days at a time.

Later in spring, when the groundswells subside, most run fish retreat farther north. That's when schools of baitfish start filling the void. Next to arrive are the summer predators: tarpon, more kings, and the last of the cobia, along with snook, jack crevalle, bonito (false albacore), and so on.

Sometime around the last week of March, schools of whitebait—thread herrings, pilchards, and Spanish sardines—start working their way up the coast. When the schools arrive, they orient to structure, which includes buoys, rock piles, and fishing piers. They turn into the current and keep feeding on plankton. Although they may not be aware of it, they're sitting ducks for the bluefish, Spanish and king mackerel, jacks, bonito, and cobia that are also migrating.

Starting in April, at Lake Worth and Juno, jacks and bonito lead the charge. Both move with the current while riding the swells, and they push schools of baitfish to where anglers await them—standing on piers armed with lures and live bait. Forty miles south in Broward County, giant tarpon are doing the same thing—minus the splashing and pyrotechnics. Then lethargy sets in with the summer doldrums, and barracudas and snook become daily mainstays with occasional migrating tarpon or stay-at-home jacks adding variety to the pier and surf fishing.

If the wind kicks up or Spanish sardines are present, there's always the chance for other species. It's not uncommon to deck a cobia, kingfish, or another spring migrant that stayed behind (or, in the case of kingfish, from another run). But this time, the weather plays a more prominent role. Bonito, for example, prefer slick-calm mornings, while kingfish hit best following afternoon showers. An abundance of baitfish—especially Spanish sardines—can spell the difference between feast and famine. Then summer winds down with the bait's departure and the onset of groundswells and northeasterly winds, and you wake up one morning with dew on your windshield and sense once again that fall's in the air.

As fall approaches, the temperature drops, imperceptibly at first, as each day grows shorter by a minute or two. That's when the sea starts changing again and migratory fish, both predatory and forage, move south. The fall baitfish migration begins well to the north; by the time it reaches the Treasure Coast, a host of predators are moving along with it.

Spanish mackerel used to arrive ahead of the mullet if the wind had been in the north. Now a few stay around all year. The surf between Hobe Sound Public Beach

Schooling mullet head south
with the start of fall Nor'easters.
PAT FORD

Bonito (false albacore) are running
on Juno Pier. PAT FORD

Hollan Hoffer holds up a 30-pound-plus jack crevalle, prior to releasing it, on the new Juno Beach Pier. Big jacks run in schools, especially during springtime. STEVE KANTNER

This bluefish hit a plug for Dr. Carl Gill at Coral Cove Park on Jupiter Island. STEVE KANTNER

and St. Lucie Inlet sometimes teems with Spanish as early as August, if there's a south-running current. Meanwhile, bluefish—a mainstay of wintertime fishing—wait for fall Nor'easters before beginning their run. Once they've arrived, they keep reappearing whenever the wind and water color are to their liking—emerald green with a south-running current. Pompano are another fall arrival in the surf as well as in the Intracoastal Waterway.

In the fall and winter, schooling Spanish mackerel, bluefish, and pompano come within reach of the beach. Fall is also when a mass migration of baitfish heads south along South Florida beaches anytime between the end of August and early November. We refer to it locally as the "mullet run." Depending on where you meet the mullet, you'll do battle with bluefish, snook, monster tarpon, giant jacks, even smoker kingfish—along with the usual sharks.

While the first major push of jacks comes south with the mullet run, their numbers skyrocket after the first of the year and continue to peak into April and May. That's when anglers fishing between Fort Pierce and Boynton Inlets encounter floating islands made up of thousands of fish that weigh up to 50 pounds. The schools, which head north just outside the surf line, travel around the piers they encounter rather than through them. Fish these schools at the new Juno Beach Pier, the surf just north of the Hobe Sound Public Beach Pavilion, and the rock-studded beach lying a mile or two south of the old Juno Pier. Jacks enter the surf in pursuit of forage, which includes everything from croakers and whiting to full-size mackerel, bluefish, and pompano. Regardless of the menu, they respond eagerly to large, fast-moving surface plugs.

Tarpon always attract the most attention. Each year during the fall mullet migration, silver kings by the hundreds pin the schools against the beach. The carnage goes on for hours. Historically, anglers have fished the run with live mullet and heavy tackle, but a growing percentage are casting plugs and, believe it or not, flies.

Run fish keep arriving through January as both the air and water continue to cool. The days grow shorter and the fish settle in. Predictable action usually follows, as the main body of pompano, bluefish, and Spanish mackerel moves back and forth between Boynton and Fort Pierce Inlets.

Then along comes March and the vernal equinox, when most migratory species start heading back north. It's as if somebody threw a switch. First, the spotted seatrout in the Indian River go on a feeding binge. Then, before you know it, the ocean migrants start leaving to wherever they go up north (sometimes as far as New England). You can find pompano through early July in Palm Beach, Martin, and St. Lucie Counties. In fact, April is the top month from Miami to Stuart.

Rampaging bluefish compete for a surface plug. PAT FORD

While most snook are small, you can catch quite a few trophies. While spinning gear is often the most practical choice, I like to fish flies whenever possible. You'll need a sturdy 9-footer—I suggest a 9- or 10-weight or possibly something longer if you intend to fish the surf. The longer rod helps keep your backcast out of the sand. A stripping basket helps keep your line from tangling. PAT FORD

Spanish mackerel show up all summer long—as far south as Juno—in the wake of a strong north wind or when schools of tiny sardines are present. Most Spanish vacate Peck Lake (a depression in the bottom a few miles south of St. Lucie Inlet) in early April, leaving commercial hook-and-liners to concentrate on kingfish. Then the arrival of June signals a return to tranquility, when the waves subside and the water regains its clarity: ideal conditions for snook.

Snook continue hitting through August; in fact, some years they'll persist into late September if the southeast wind keeps puffing away. Game and fish laws are constantly changing, though, so before keeping a snook, check the regulations (www.myfwc.com). Snook thin out as fall approaches, an event that signals the next mullet migration.

Equipment Essentials

While saltwater gear works fine in fresh water—as well as in brackish—the reverse isn't always the case. The salt imposes additional demands, based on tougher fish and corrosion. The fish not only grow larger here; they're also much stronger.

Whether I'm fishing a pier or the surf, or just wading and casting, spinning gear makes it easier to cast small to midsize lures or baits. Anything weighing 5 ounces

Jeannie Eastman shows off a
snook. Gamefish come closer
to shore at the top of high tide.
PAT FORD

Modern-day spin gear and super-
braid line offer an advantage to
anglers when big fish are running.
PAT FORD

or less is easier to handle with a fixed spool or spinning reel; if it's heavier, I switch to a conventional model. Spinning rod action tends to be faster than a conventional rod. It produces a snappier send-off—one more like a launch—that's ideally suited for casting lures. Think longer, with a more flexible tip.

When purchasing a spinning reel for the salt, I first look at line capacity. Since I fished from piers in my formative years—where big fish run—line capacity was always important. You can't follow your fish from a pier, after all. I also want a reel with a fast retrieve, but not too fast. I want my spinning reels to have at least a 5-to-1 ratio. While a smooth drag ranks high on my list of requirements, I can usually improve mine by buffing the washers. Shimano's Stradics fill the bill for me.

Other features have now become standard, from one-piece spools (which are less likely to warp) to corrosion resistance. Even the bails have been strengthened to help to prevent bending. Modern spinning reels feature ball bearings to ensure smooth operation, but if you want additional smoothness, reach for your wallet. That's why manufacturers offer similar models with different price tags. Before selecting a reel, I ask make sure I can grasp the handle between my thumb and forefinger. Something else I always look for is a free-wheeling line roller with a large bearing surface. That makes a major difference with a fast-running fish.

Some reels come equipped with a manual bail, which allows you to remove and replace the line with your forefinger. Keep in mind, however, that this takes some practice. Since there's less to go wrong, I salute the concept. Still, most reels I see feature cam-loaded full bails—some equipped with more than one bail spring.

An angler on a beach free-spools a big-game reel while a kayaker carries his bait offshore.
STEVE KANTNER

Lures are expensive these days, so here's a tip for holding onto yours: Many spin fishermen allow their bails to flip and their manual rollers grab the line while casting. You can break fish off and lose lures that way. Be sure your spool is fully extended before attempting to cast. (This keeps your line beyond the reach of most moving parts.) And move your handle to the downward position, away from the reel foot, so that it doesn't rotate.

The conventional reels that I use (reels where the spool revolves) run the gamut from lightweight to truly large. The pair of Garcia Ambassadeurs that I use in the surf are engineered from aluminum to extremely close tolerances, which keeps the weight down and sand from the gears. Meanwhile, most midsize conventional reels are larger and tougher and hold between 250 and 450 yards of 30-pound mono. You may need it on piers, where the Gulf Stream's the limit. I use mine on piers and bridges as well as in the surf—especially during the mullet run. Most who fish with live bait on bridges fill theirs with 60-pound mono.

I own several big-game reels, which depending on their size, hold 300 or more yards of 50-pound mono. One's a Penn 4/0; another's a 6/0 that I use for ballooning baits. You'll want plenty of capacity for swimming out live baits or landing large sharks. I still have an old 9/0 Penn Senator that holds 400 yards of 80-pound mono, just in case. All my shore fishing reels have standard star drags, although most new ones I see come equipped with a lever, which provides more even pressure thanks to a larger drag surface. Both Penn and Shimano market excellent examples.

Spinning Rods vs. Conventional Rods

The most important things to look for when choosing a spinning rod are stiffness and action. Meanwhile, thanks to today's faster tapers, the same stick will perform well with a wide range of line tests—which are typically marked on the blank near the fore grip. Length is another, albeit separate issue, with most spinning rods—not including those built exclusively for surf fishing—measuring somewhere between 6 and 7½ feet. Some surf sticks, remember, are stiffer than others. Ranking next, when it comes to choosing, are quality components: meaning the blank, the reel seat, the guides, and the fittings. If it's truly a spinner, does it have the right guides?

Most conventional rods—we're taking pier and bridge and not boat rods here—are built on fiberglass blanks that measure from 8 to 10 feet in length: thick-walled shafts that may weigh a pound or more. Carbon fiber versions are lighter but more expensive. Whether they're custom-made in a shop or come fresh off the rack, be sure they're sufficiently stout to handle your line, bait, and sinker. You'll learn to judge as you gain experience. But for now, you'll have to rely on the description that appears on the blank—or the word of the salesman.

Perhaps the greatest difference between conventional and spinning rods—next to the rod guides—comes down to action. Whereas spin rods tend to be tippy, most conventional sticks are more parabolic, making them slower and less likely to create a backlash.

While used reels can be reconditioned—offering a substantial savings to customers—that's not the case with rods. You'll never know for sure what abuse they

Anglers who fish bait for everything from pompano to oversize kingfish are the primary proponents of conventional tackle. This Ward Woodruff custom rod sits in a sand spike.

MIKE CONNER

sustained and about nicks and dings that may lie hidden beneath the finish. While that's not always the case, use caution when purchasing a pre-owned rod.

Plug Gear

You won't see plug gear used as much in the salt as in fresh water—or at least, not as often. One exception is where anglers cast heavy lures: weighted grubs from a jetty, or plugs or jigs below a spillway. Success in these venues is based less on distance than on control—hence the popularity of plug over spinning gear. Saltwater plug casters prefer reels with a star drag such as those from Penn and Shimano. They match them with 7- to 9-foot rods and lines testing anywhere from 15 to 50 pounds. You read that right. Think snook in a crowd.

Surf Gear

Until recently, the standard for fishing the breakers, at least if you're using bait, consisted of a one-piece, 14-foot rod with medium to heavy action. It was fitted with conventional guides, matched with a lightweight, corrosion-resistant conventional reel engineered to extremely close tolerances in order to keep out the sand and the salt. The Garcia Ambassadeur series—notably, models 7000 through 10000—represent the epitome of this genre.

Surf fishermen who purchase 7000s, which are excellent casting reels but are constrained by a level wind mechanism, sometimes take off the reel's right-hand side plate and remove both the worm and level wind gears. I did this myself to

Saltwater Wading

Wading anglers can also target bonefish in Biscayne Bay, within sight of downtown Miami. I've done plenty of wading in this body of water, where catching bonefish is no longer a given. However, the ones we land there are large on average, in excess of 7 pounds.

According to Captain Carl Ball who knows Biscayne Bay like the back of his hand, bonefish come within reach of waders at Matheson Hammock Park (on Old Cutler Road in Coral Gables) where anglers can access the bay by walking off the bank to the south. He mentioned wadable shoreline all the way to the mouth of the Snapper Creek Canal (C-2), just north of Shoal Point, as well as near a spillway that lies west of Chicken Key, where the bottom's hard and where bonefishing can be good.

Carl also says that you can fish Key Biscayne proper on the ocean side out of Crandon Park. Either walk the beach or wade the flats. At the south end of the island, you'll find Bill Baggs State Park. While the park is worth exploring, this type of fishing depends on tide—both its height and direction—since bonefish follow a predictable path both onto and off each flat. So you're better off covering lots of water. To get the most bang for your buck, I suggest you hire a skiff guide like Carl.

The man behind the legend: D.O.A.'s Mark Nichols admires a seatrout. PAT FORD

Captain Ed Zyak with a spotted seatrout. PAT FORD

In my opinion, the greater potential for saltwater wading lies north along the Treasure Coast in the Indian River Lagoon (IRL) and parts of Jupiter Sound, where you can catch spotted seatrout, redfish, and other species—and access is very good. According to Mike Conner, "The redfishing now is some of the best I've seen since the late '90s, and these fish are accessible to waders. Just look for well-worn paths at the side of the beach road (A-1-A), and follow them down to the river."

Anglers catch reds and trout, along with the ubiquitous ladyfish, on live shrimp, as well as an assortment of lures. Among the latter is the D.O.A. Shrimp, a weighted soft plastic with lifelike appeal.

The best fishing usually takes place near the Hutchinson Island Nuclear Plant, where anglers walk in from A-1-A. This area is popular with kayakers, too. I have yet to try my canoe due to the wakes from passing yachts.

There's more wade-in fishing south of Hutchinson Island; namely, in Jupiter Sound. I frequently see waders near the Hobe Sound National Wildlife Refuge Headquarters— located just off U.S. 1 in Hobe Sound. To reach the fishing, you descend from a bluff and

(continued on page 160)

(continued from page 159)

Redfish populations are on the rise after several major cold fronts leveled their playing field by killing off snook. PAT FORD

follow a path to the water. Either wade sideways or east toward the channel. Portions of these flats—especially near the mangroves—harbor small snook during winter.

Drive on A-1-A along Hutchinson or Jupiter Islands. You'll see plenty of places where it's legal to pull over and where the shallows extend from shore. Access paths are easy to see. Look for docks while you're at it; they usually attract bait. In addition to the species I previously mentioned, most flats in this area host flounder, bluefish, and pompano—depending on when and where you look.

enhance the reel's casting ability. Pompano fishermen are famous for tricking-out tackle. On the other hand, stock equipment has been making inroads into the custom market. I currently rely on several two-piece models marketed under the name Tsunami, ranging in length from 10 to 12 feet.

I also have surf spin sticks, also—again Tsunamis—on which I've mounted a pair of 8000 Shimano Stradics. That's what I use for bluefish and spinner sharks, and whatever else that will hit a heavy lure. I use the 10 and 11-footer for casting hollow surface plugs that I fill with BBs, so they weigh 5 or 6 ounces. I can cast them a country mile.

Rods used in the surf can also be smaller. I spend entire days casting tiny jigs with 6- to 6½-foot medium-action spin rods matched with tiny reels filled with 6-pound-test monofilament or its stronger super-braid equivalent. One of my favorite outfits is a Shimano 2000 filled with 15-pound PowerPro and a 6-foot, fast-action rod I purchased at Bass Pro Shops. Another, I spooled with 8-pound-test, which has the approximate diameter of 2-pound mono. That makes it easier to cast, as well as letting you get more on the reel.

There's middle ground, too. For making long casts with spoons—like I do on the piers—I prefer a heavier-action, 7-foot spin rod and a slightly larger Shimano 4000 reel that I fill to the rim with 10-pound mono. The mono shows up well in the air, which helps avoid collisions with seabirds. Plus, I like stretch when I'm fishing a spoon, rather than having to dodge it when a fish spits it back at me.

Big Fish Tackle

Literally millions of mullet migrate south along South Florida beaches during the fall baitfish migration. While some anglers are content to catch smaller predators—Spanish mackerel, bluefish, and the ubiquitous ladyfish—a few hardier souls yearn for something more. They are after the big fish that follow the mullet schools—everything from tarpon to oversize kingfish that come within reach of the surf.

These big-game anglers fish either a live mullet or a fresh mullet head. A typical setup consists of an 8- to 10-foot rod and a conventional reel that's capable of holding at least 300 yards of 40-pound mono. The leaders are heavy—at least 100-pound mono—and the hooks are strong: a flat-forged 8/0 is just about right.

Other big-game anglers tackle monster bull sharks, hammerheads, tigers, and more by rowing out large baits from the beach—say, a whole dead bonito. This is discussed in detail in the surf-fishing chapter, but here's a synopsis of the gear that's involved.

Let's start with the reel: you'll need a high-capacity, big-game model that's capable of holding at least 500 yards of 80-pound mono, or its thinner but more expensive Dacron or Micron equivalent. Then there's the rod, which is typically longer than you'd use in a boat (it helps keep the line from the breakers). Next, you'll need a kayak or rowboat (a kayak is safer) for rowing out baits and a flashlight for signaling. Add to this some super-strong leader material (at least #12 wire) and heavy hooks (size 12/0 is popular). Plus, you'll need something to hold your bait on the bottom, once you drop it off. Half a concrete block excels in this capacity.

Saltwater Fly Gear

Fly Rods

You'll need a sturdy 9-footer—I suggest a 9- or 10-weight—possibly something longer if you fish the surf. The longer rod helps keep your back cast out of the sand. The nine's also my choice for wading for seatrout, due to the weighted flies.

I use lighter rods where wind isn't a problem. In fact, they're easier to cast once you learn to build up line speed, which is mostly a matter of loop control. A perfect description of Seven-Weight Heaven is a glassed-off surf when the snook are swimming.

Fly Reels

Fly reels for the salt come in two basic versions: lightweight with fiber or caliper drags and those "Big Kahunas" with brakes like a jetliner that feature a cork disk brake pad. These beefed-up models have one-piece frames that are milled from aircraft-grade aluminum. You don't need a bar-stock reel for everyday fishing, but they can make a big difference if you target the monsters: say, tarpon in the mullet run, or spinner sharks. As far as my favorite brand? Tibor, regardless of size.

If you are planning on hooking a monster, purchase a reel with sufficient capacity, at least 300 yards of 30-pound backing, plus an 11 or 12-weight line. Hook a spinner and he'll take that and more. For rank-and-file fishing, I try to keep it simple: a light-weight reel filled with at least 150 yards of 20-pound backing. The standard are 7- and 8-weights.

For mackerel, blues, or snook in the surf, I prefer lower-priced models with holes in their side plates—to let the sand in and out—or something more expensive that's machined to closer tolerances. You don't need much drag. Tibor's Lightweight Series is a perfect example.

Fly Lines

Use weight-forward or saltwater taper fly lines, with the proper sink rate for what you're doing. Floating lines, as usual, are always appropriate; however, clear "slime lines" and sink-tips score bigger in the surf. Full-sinkers, if you have them, work best in an inlet.

Other Essentials

A number of outfits carry hook files in their catalogs, but for serious work I head to the hardware store. I take a flat bastard file about six inches long and coat it with grease or WD-40 before stuffing it back in the scabbard it came in. A rusty file is totally useless. As a safety precaution, it stays in my tackle box (or bag).

Here in the salt, sunglasses are a necessity, not just for reducing glare and spotting fish, but for preventing injury. My personal choice is Maui Jim. They're state of the art and comfortable, too. I prefer gray lenses for the ocean and yellow or brown in the boonies.

Make sure to carry water to drink. We absorb cold water better than warm, so I always keep extra water on ice. Adequate hydration is especially important if you

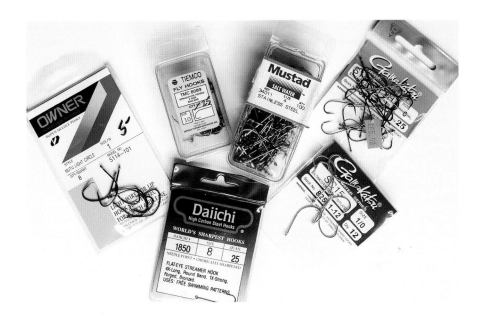

Hooks these days come in every conceivable pattern. Designer models, while they may seem expensive, are worth the price for specific tasks. I still use basic J-hooks—like Mustad 9174s with live baits—although they've been replaced by newer circle models. But I bought them in bulk, and I injure few fish. A good hook hone is essential, as well as a tool for removing hooks from toothy bluefish and mackerel—I like the Hook-Out. STEVE KANTNER

Appropriate saltwater attire for warm weather includes a light-colored, breathable shirt, hat, sunglasses, and quick-dry shorts or wading pants. Here, a wading angler hefts gator seatrout. That's guide Ed Zyak of Stuart. PAT FORD

take medication. My surgeon friend taught me to hydrate with G2 Gatorade during hot weather. It's an electrolyte-replacement drink that helps maintain blood volume better than water does.

Your physical condition can be more important than your equipment. Saltwater fishing, more so than its less briny equivalents, can be serious business. Competition is stiffer and the fish grow larger—large enough to wear you down. Frolics in the brine can be rugged encounters, and the physical demands are much greater than in fresh water.

Learn to cast correctly before you head out, rather than endangering yourself or others around you. Then learn to reel correctly by grasping the handle firmly between your thumb and forefinger. Forget about using the palm of your hand—that inhibits your ability to retrieve line quickly.

Learn important techniques as fast as you can: wrapping wire and splicing mono to braid. You can catch on quickly by reading the packages or from a well-illustrated book (such as *Fishing Knots* by Lefty Kreh).

There's an intensity about the salt that you won't find elsewhere. Try to stay focused, but if you need a break, retreat from the crowd before lowering your guard or removing your eyewear. Getting hooked is just one glaring example of what can go wrong; there are other less obvious ways to get into trouble.

Sleep at the switch and you'll limit your catch, too, especially on piers and jetties, where there's competition.

Strategies are part of everyone's game, and no doubt you'll develop a few. In the meantime, observe other anglers and strive to perfect your basic technique while keeping your tackle in tip-top shape.

South Florida Surf Fishing

Surf fishing isn't for the faint of heart. If you're lucky, you have access to a private beach—even one where floodlights attract baitfish and predators. But the rest of us have to access surf spots by occasionally slogging through miles of sand—which becomes increasingly difficult as the tide rises, due to the softer sand above the high-water mark.

There's something primal about hiking for miles on end on a desolate beach (they still exist here, thanks to wealth and privilege) and encountering a school of 40-pound jacks or other quarry. PAT FORD

Long walks in the sun for fishing that's unpredictable are just part of the game to a cadre of veterans who believe that "one from the beach is worth ten from the boat." Nothing compares to the thrill of working a fish through the breakers and dragging it up on the sand—it's just you and the fish. The surf offers excellent fly-fishing opportunities, too, from snook in summer to bluefish in winter. The long rod is also practical when Spanish mackerel, jacks, and ladyfish invade the surf chasing tiny baitfish.

A growing percentage of surf casters these days go just to escape their routines for hours or days at a time. They're retirees, mostly, who meet on the beach to toss out some bait and enjoy the scenery. They may end up catching a limit of pompano, or maybe a dozen blues. But what they're really doing is passing the time—casting and reeling, while enjoying the day. While that might be enough for the casual enthusiast, there's a second, more committed approach.

Anglers in the second group target particular species by employing specific methods or with special baits or lures. The majority show up during established seasons, at specific times of day, or during particular tidal phases. Although nothing in fishing is certain, deliberate strategies improve their chances.

Beaches, where possibility rides every wave, create their own brand of magic. Who can predict when a king or a tarpon will hit? Or when a permit will discover your sand flea rig? Mike Conner caught four nice permit from the surf in a single day. Avoid beaches where the water's exceedingly shallow; it's the surf, after all, not a bonefish flat. And try not to arrive as the tide bottoms out. The odds are in your favor during the last three hours of incoming tide through the first two of the outgoing. STEVE KANTNER

During South Florida's fabled mullet run, schools of baitfish head south along the beach. PAT FORD

Fall Baitfish Migration

A major draw every fall is the baitfish migration that coincides with the onset of northeast winds. At one time, schools of mullet of assorted sizes stretched for miles and included every conceivable whitebait species (pilchards, thread herring, Spanish sardines, etc.). You can only imagine the stir it created. Even today, it attracts its share of predators.

You'll still find plenty of beef in the breakers. That's why heavy gear for the mullet run consists of a heavy-duty, 10-foot surf rod matched with a conventional reel capable of holding at least 300 yards of 40-pound mono. The bait of choice here is either a live mullet or a fresh mullet head, either free-lined or fished on the bottom. Size 6/0 to 8/0 forged ring-eye hooks are standard fare, as is a 3-foot mono leader—testing between 100 and 150 pounds. Adding a short length of wire is optional.

While tarpon and snook are the primary targets, monster jacks, spinner sharks, and occasionally, king mackerel enter the shallows when mullet are migrating. One day last fall, two 20-pound kings were landed at Hobe Sound Public Beach. Both hit lures while the mullet were running. The majority of anglers who fish the mullet also carry lightweight spinning rigs, like the ones you would use on a pier. That allows them to target those lesser predators that gorge on the smaller forage that accompanies the mullet: tiny finger mullet, glass minnows (bay anchovies), and rain-bait-size pilchards.

You'll also find Spanish mackerel, bluefish, pompano (which occasionally feed on baitfish), and ladyfish—many of which are bragging-size. Also jacks, blue runners, moonfish, and ribbonfish (cutlassfish) that some anglers consider nuisances. Wobbling spoons, along with plugs, nylon jigs, and marabou crappie jigs, are standard fare

Spanish mackerel invade the surf in pursuit of glass minnows and other small baitfish. PAT FORD

on the beach at this time, although fly fishermen have recently gotten in on the act by casting streamers and occasionally saltwater poppers.

Run-style fishing continues well into winter, when the wind and surf enjoy a respite or when west winds prevail. That's when migratory species take up residence just offshore and return to the breakers when the conditions are right. Then the weather takes a turn by the middle of March, and the migrants depart in the wake of southeasterlies—leaving whitebait schools to fill the void. The arrival of June signals a return to tranquility, when the waves lose their punch and the surf flattens out—prime conditions for snook.

Major Surf Species and Tactics

While inland fishermen keep searching for access, surf casters stay busy with seasonal run fish. They also encounter schools of small jacks, along with the usual blue runners and sometimes ladyfish—especially during fall when the run fish arrive. Giant jacks steal the stage during spring between Boynton and Fort Pierce Inlets.

Snook and tarpon show up in the surf as soon as the water warms. Schooling sharks are there, too, from fall through late spring. However, large individuals don't follow a schedule—other than responding to the presence of prey. So use caution whenever you swim or wade and heed all posted lifeguard warnings.

An array of different species inhabits the surf, but most avid surf fishers target only a few, starting with pompano, bluefish, jacks, and occasionally sharks. Others may concentrate on pan-size nibblers like croakers and whiting, while a few target Spanish mackerel—which you find more of around piers.

Pompano

Both surf and pier fishermen recognize specific locations where, year after year, schools of pompano run. Anglin's Pier in Lauderdale-by-the-Sea, the green-roofed condo on Juno Beach, Loggerhead Park south of Juno Pier, and Hobe Sound Public Beach: all four are popular with hook-and-liners who pursue fish with a passion.

These fish appear to migrate, but not always along the beach. I say that because April is always a stellar month throughout the pompano's range. It's almost as if they swim inshore from somewhere, possibly to spawn in the Intracoastal before fattening up in the surf.

A typical setup for pompano includes a 10- to 14-foot surf rod matched with a spinning or conventional reel that's capable of holding at least 250 yards of 20-pound mono. You'll need the extra line if you hook a big permit or shark. Plus, the larger reel makes casting and retrieving easier. Sand spikes and flea rakes are other necessities, as is a serviceable cooler. If you can't find sand fleas, purchase clams or

Pompano: the piece de resistance of gourmet (as well as commercial) fishermen. Mike Conner and the author tap Hutchinson Island. MIKE CONNER

While live or dead sand fleas (mole crabs), fresh dead shrimp, and pieces of clam are standard bait for pompano, you can also catch them on lures such as Doc's Goofy Jig. The best time to stock up on sand fleas is during September. MIKE CONNER

the freshest available dead shrimp. Clams will keep in a freezer for a year if you clean them, cut them into thumbnail-size pieces or slightly longer strips, and layer them in a container packed with salt between each layer.

I cast my pompano rigs slightly upwind, which helps my sinker stay anchored and keeps my line clear of pesky seaweed. As the rip current pushes away from the beach, it carries with it forage that attracts the pompano. I attempt to cast just beyond the edge—where the sand meets the clearer water—and watch my rod tip for signs of activity.

I look for a bump or any indication that a pompano has picked up my bait. Slack line is another clue. The Kahle hooks I use (sizes 1 through 2/0) set themselves. But rather than relying on a passive approach, I frequently check my baits or stay busy by casting a lure.

A typical pompano rig consists of two or three Kahle hooks rigged in series above a pyramid sinker—preferably a four-sided one that won't track to the side whenever you attempt to retrieve it. I slip my hooks over 3-inch loops after crimping the loops between my thumb and forefinger and forcing them through the hook eye. That way, I can exchange them when they become dull or rusty. Some anglers adorn their loops with Styrofoam balls or red plastic beads; others are content to fish theirs bare. Gold hooks are another option.

When my interest in surf fishing became more consuming, I decided to purchase a beach cart. It's a Fish-N-Mate Jr. and it makes hauling gear easier. If you have money to spare, buy the big balloon tires. Other essentials include sand spikes and a flea rake. STEVE KANTNER

Goofy jigs share space with skirted jigs and ⅝-ounce Krocodile spoons in a handy packet. Goofy Jigs are ice fishing lures that are deadly on pompano. Simply toss one out and let it hit bottom before giving your rod tip a snap. Then let the Goofy Jig sink and repeat the process, after recovering the slack. Fish them on 15-pound-test PowerPro to avoid breaking them off. STEVE KANTNER

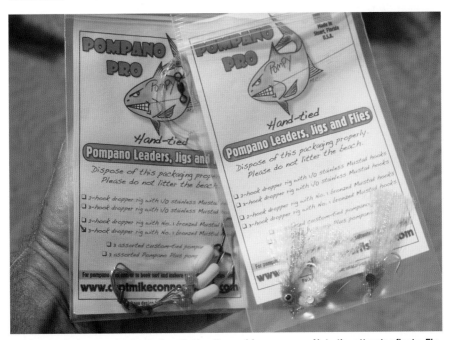

Multiple hook rigs are standard when fishing the surf for pompano. Note the attractor floats. Fly fishermen pursue them with brightly colored jig flies in the Indian River. MIKE CONNER PHOTO

Bluefish

The surf fisherman's default quarry is the bluefish, especially north of Palm Beach Inlet, where the runs are consistent. I pursue them there with oversize surface plugs that perform yeoman service with the 4- to 10-pounders that run the surf between Stuart and Palm Beach Inlets. Look for blues at Hobe Sound Public Beach as well as a mile or two south of the old Juno Pier from November to March. Bluefish bite best on a north or northwest wind.

Smaller bluefish hit big plugs, too, but often tear off while you're reeling them in. A better bet for them is a smaller plug: try a MirrOlure Top Dog Jr. loaded with BBs or, my favorite, the X-Rap Rapala in size 10. Of course, you can still fish for blues the way I learned to—with a Krocodile or Kastmaster spoon.

Head south and you'll see how the runs have dwindled. Schools of winter-resident, 4- to 7-pound blues seldom make it south of Boynton Inlet like they did in the past.

They still, however, enter the Intracoastal Waterway through Palm Beach, Jupiter, and St. Lucie Inlets. Although bluefish—along with jacks, ladyfish, pompano, and occasionally, Spanish mackerel—are caught from seawalls in the Lake Worth Lagoon (especially in the vicinity of West Palm Beach's Trump Plaza), the fishing there is now less predictable.

Even farther south, in Broward and Miami-Dade Counties, we experience runs of Jersey blues—those 13- to 20-pound hollow-bellied monsters that visit South Florida during the spring, to spawn. While they typically show up around the middle of March in 400 to 600 feet of water, a strong Nor'easter will push them ashore. These saltwater piranhas hit anything that moves—including bathers during one infamous blitz.

Bluefish lead the list of surf-fishing targets. PAT FORD

A large expanse of seawall that's accessible to the public on the western shore of Lake Worth Lagoon. STEVE KANTNER

One guy showed me where he'd lost a finger while swimming alongside Pompano Pier. But like all daytime surf fishing along the entire Gold Coast, just getting to the beach is a major problem during the day. Then pier fishing is a better choice.

Gold Coast anglers see plenty of bluefish—albeit, small ones—first in October and again in March. These fish average between 1 and 3 pounds, and their weight increases as the season wears on. We get several days during the mullet run when the schools swim close to the beach.

Due to laws that govern surf fishing, which vary according to jurisdiction, blues are best pursued from piers. And yes, they hit at night—typically on bait just after sundown. Anglin's in Lauderdale-by-the-Sea is known for its bluefish, as are Dania, Pompano, and Deerfield Piers, depending on surf conditions and where the run is. Pier fishermen in Broward look for fall Nor'easters and spring southeasters—the latter starting around St. Patrick's Day.

Some piers are surrounded by buoys positioned to keep boats and skin divers away. That's no place to hook fish that make long, sustained runs.

South of Port Everglades, the run starts either earlier or later, depending on whether it's spring or fall. While pier fishermen in Broward encounter more mackerel than their surf-fishing brethren, that's not always true farther north.

Bluefish generally prefer something larger: a 3/0 streamer tied from synthetic materials, or when conditions are right, a saltwater popper. A short shock leader fashioned from No. 2 single-strand wire helps prevent cutoffs. While a 9- or 10-weight outfit is standard in the surf, fly fishermen occasionally hook truly large gamefish, notably during the mullet migration.

Mike Conner showed me a newly tied pattern that closely resembled a pilchard—pink highlights and all. When I asked what he called it, he replied, "Just a Clouser," before landing fifteen snook to my two or three. He ties it with Ultrahair on a 2/0 hook. MIKE CONNER

You'll need a slow-sinking line (preferably a low-visibility model, such as Scientific Anglers Striper Line) and a selection of streamers. If you're rigged with a floater, don't forget poppers. A Lefty's Deceiver is hard to beat, especially a white one in size 4 to 2/0.

If the fish are on glass minnows or tiny pilchards, then a sparsely dressed bucktail or Glass Minnow pattern is made to order. Personally, I think I've improved on those patterns with these modifications: I tie a tiny Deceiver on a size 6 hook with four white saddles surrounded by a pair of grizzlies. To this I add a sparse calftail collar and painted eyes—the eyes more for me than my quarry. Occasionally, I include a thin strip of flash. Some days it works better than others, but there's seldom a time when the fish won't hit it.

Snook

Fly fishers in particular take advantage of these linesiders that ply the breakers from May through the fall. Anglers also target snook with live bait such as live mullet. When surf fishing for snook with a live bait, I prefer to anchor it, which typically requires at least a 4-ounce sinker. For tarpon, I sometimes free-line. Snook also hit mullet heads, especially fresh ones, during the summer when they search for sea turtle hatchlings.

Guide Ed Zyak lifts a fly-caught snook. PAT FORD

If whitebaits are present, then that's what I use, rigging a live one below a sliding sinker. I've even seen anglers hook up two tiny live pilchards at a time, both through the nostrils. Other local favorites include croakers and whiting—especially in August. If you can't get either, snook will also hit pinfish.

When it comes to lures, a jig's hard to beat: either a lightweight white bucktail or a Red-Tailed Hawk, which might resemble a whiting. I reel both diagonally to the breakers. Other lures that are popular include a varied assortment of grubs and jerk baits—the former I usually fish on a lightweight jighead. I prefer the more neutral shades, ideally camouflaged against the white sand bottom.

Jacks

Jack crevalles remain one of my all-time favorite fish to catch, especially those 25- to 50-pound monsters that roam the surf from February through May. An inshore reef forms a natural funnel south of Hobe Sound Public Beach through which these leviathans make regular incursions before running all the way to the beach. You'll find another hot spot north of Lost Tree Village, just south of the town of Juno Beach.

A few big jacks arrive with the mullet run, but wait until after New Year's if you're looking for action. If the water's cold, the fish may be listless and refuse to hit lures. This may also be the result of spawning activity or too much pressure from anglers in boats. The schools don't come as close to the beach at this time, preferring instead to raise anglers' heart rates on piers. But by the middle of March, or when baitfish arrive, all that changes and the bacchanal begins.

Jacks slip down the swells in angry battalions, driving all manner of baitfish to eventual oblivion. That's the time for lures, especially surface plugs, and few perform better than a customized Lazer-Eye filled with BBs to add weight for casting.

Smaller jacks range the beaches, frequently in massive schools, frothing the surface and devouring tiny forage. The best time for schoolies is from September into March. They, like bluefish, hit practically anything if it moves fast enough, including flies and poppers.

Spanish Mackerel

Both Spanish and small king mackerel are equally addicted to tiny forage. Sometimes you have to use heavy surf gear, if that's all you're carrying—although a light rod and spoon is preferred.

If you're stuck with that full-size rod, combine a full-size plug, a clear plastic float filled with seawater, or a wooden dowel with a trailer, which might consist of nothing more than a 4-foot length of 40-pound mono to which you've attached a Clarkspoon or jig.

I make mine more elaborate by adding a string of nylon quills tied on stainless hooks (Mustad 34007). Three's the magic number for quills: two and they ignore it; four and it tangles. Work your float as you would a popper. This is a good rig if you're only toting one rod and it's too large for casting a lighter lure.

Smaller Surf Species

A variety of small fish inhabit the surf—several of which are pursued for their culinary value. Croakers, for example, become locally abundant in the summer. You can spot the schools at the edge of the breakers. Snook, of course, have figured that out, as have the jacks and bluefish that slice through the beakers.

Both yellowfin croakers and Norfolk spots—common varieties along the Treasure Coast—prefer deeper water than whiting. You'll find croakers near shore on Lake Worth and Juno Piers, suspended over the white sandy bottom, where their shadows stand out.

Whiting, on the other hand, prefer the curl of the surf. They're excellent table fare—in my opinion, outranking croakers in both fighting ability and taste. And they don't have as many worms.

An effective way to target both croakers and whiting is with a small piece of freshly killed shrimp, or a tiny sand flea, fished on a size 8 hook on a sliding sinker rig. A ½- or ¾-ounce egg sinker should be enough. For croakers, allow the rig to sit on the bottom; for whiting, let it roll with the waves. The fish hook themselves, so hold on tight. Fresh shrimp, like I said, works better than frozen. Bloodworms are also effective for whiting, as are small crappie jigs.

Like to cast that jig farther? Try slipping a shell over the hook. I learned this tip from Terry Luneke, who taught me how to catch whiting on marabou crappie jigs. If you pick a shell with a big enough hole, when you jiggle your rod tip it falls off the hook. When Terry lands a whiting, he rubs his marabou jig through the slime to put the scent on the chenille and fibers. STEVE KANTNER

Permit

Other novelties that ride the curl include schools of tiny permit, especially in Broward County during the summer. I catch them accidentally in my sand flea rakes, and I assume they get here by following the current from their birthplace in the Keys.

Catching adult permit is another matter, but you can land them on sand flea rigs. Permit come big and provide a challenge—by big, I mean 30 pounds and up. Most anglers release the larger ones, although the law doesn't say that you must. While the majority of permit are landed in Broward and Palm Beach Counties (typically from piers), these fish are fairly common as far north as Sebastian Inlet.

Sharks

Spinner sharks (*Carcharhinus brevipinna*) are becoming more popular with sport fishermen in Palm Beach and Martin Counties, where skiff guides pursue them with flies and light tackle.

Captains Scott Hamilton, Butch Constable, and Ron Doerr have brought this exciting sport to its present level. The sharks appear with the mullet; by the end of January, their ranks have swollen so that surf casters have daily encounters with them, which isn't that cool if you're reeling in pompano. When the spinners invade

Maggie Luneke releases a spinner shark. PAT FORD

the surf in numbers—usually around high tide—it's nearly impossible to land any-thing else. But they blast a surface plug like few other species and give you the fight of your life.

At their worst, sharks are dangerous predators that attack human beings (I know one victim). Surfers are a favorite target. But from the relative safety of a sandy beachhead, they're a line-ripping jumper that will outfight a billfish. Spinners have a reputation for breaking off rigs while performing their signature leaps—the angler's line wraps around their skin or catches their fins, both of which are covered with dermal denticles. These shark scales are essentially tiny teeth that sever most lines on contact.

I've been partially able to circumvent the problem by spooling my Shimano 8000 with 50-pound PowerPro, the last 15 feet of which I double. While no spin-ning reel drag that I know of can exert enough pressure to break 50-pound line, the abrasion resistance afforded by the PowerPro is an effective deterrent to the con-stant cutoffs. And since a hooked spinner can run off 300 yards at a clip, it's com-forting to know that the reel can hold it.

I attach a medium-size black snap swivel to my doubled line by slipping the loop a few inches through my swivel before opening either side and spinning the swivel through both sections a total of four or five times. I call this a Japanese Bimini, and it looks a lot like double clinch knots.

I use a heavy mono leader (2 feet of 150-pound Ande) with a surgeon's loop in one end so I can easily join it to my snap swivel. My plug is tied directly to the mono, but if I intend to recover it, I debarb the hooks. Any attempt to unhook a spinner is dangerous, though. I always use a long piece of driftwood.

If I'm fishing bait—say, a chunk of blue runner or bluefish—instead of a lure, I substitute a 5-foot length of 150-pound-test mono to which I've added a 12-inch piece of No. 12 wire by splicing the two with an Albright knot. The hooks I use vary between sizes 5/0 and 8/0. Pinching down the barb helps the shark expel it after I reach down and unsnap the leader. This part is tricky—and dangerous, too.

While 100-pound spinner sharks hit lures or baits cast from shore, most of the trophy hammerheads, bulls, and tiger sharks landed in the surf on heavy tackle were caught on baits that were carried offshore and dropped from kayaks. I've heard from friends who used to fish the area that the Delray Beach City Commission has outlawed the practice. So check all laws before attempting it.

In the 1960s, Lake Worth hardware salesman Herb Goodman was featured on national television for his exploits with sharks. The show, as I recall, was "I've Got a Secret," although "What's My Line?" might be more appropriate. Herb fished the north jetty at Boynton Inlet, where he beached dozens of monsters—not only by ballooning out (see the next page), but by sending out baits in a radio-controlled boat. Herb or one of his cronies would rig a large, dead bait—say, an entire bonito carcass—on two or three 12/0 or 14/0 flat-forged Pflueger Martu or Sobey hooks, which he'd attached to a 20-foot cable or twisted No. 15 piano wire leader. He tied balloons to the snap swivel and cast the bait into the current.

Giant bull sharks, a species known to attack humans, are frequently landed by youthful shark-ers—some who fish from the beach. CHRIS BISHOP

Using a kayak to carry out shark baits. STEVE KANTNER

Heavy tackle is a prerequisite for monster fishing, meaning at least a 9/0 reel filled to the bars with 80-pound mono or Dacron. A 12/0 or 14/0 reel is even better. The trick is to drift your bait to the strike zone—500 or 600 yards offshore—before rearing back and popping the balloons. The bait will then sink to the bottom and, hopefully, into the jaws of a hungry shark.

Unlike on a pier, you're using current. To take maximum advantage of the outgoing tide, allow your bait to dangle deep in the water without snagging bottom.

Because of difficulties inherent in casting bait rigged in this manner (Try casting a 10-pound bait at the end of a 15-foot leader!), some anglers strip off just enough line from their reels to get their baits to the current and then swing bait and leader overhead like a lariat. Be careful not to lose your balloons if you try it. Use small oblong balloons so they won't be affected by wind. Tie them to your snap swivel before tearing off the nipples. Increasing boat traffic has been an obstacle to ballooning baits out of inlets, although boat traffic varies according to sea conditions.

These days, anglers use kayaks to carry out shark baits. Sharks follow specific depth curves, and your odds improve several hundred yards from shore. The kayaker drops a double- or triple-hooked bonito or other large fish carcass at least 300 yards from the beach or waits for a signal from land. Beach fishermen, typically, prefer larger baits because that selects against smaller predators, such as rays and barracudas. Again, truly large reels—size 9/0 and better—are standard equipment for this approach.

Anglers who fish for sharks from the beach attach chunks of cinderblock to their baits with heavy mono. That keeps their baits from rolling with the current. When a shark eats the bait, the cinderblock sinker separates, leaving the angler tight to the fish.

Surf Spots

Years of development have left too few places to access the ocean—unless you live in a beachfront condo or know someone who does. Those few exceptions are public beaches. Anyone who's serious about fishing the surf should head north to Palm Beach, Martin, or St. Lucie Counties, where the citizenry cares more about nature than it does about maintaining a tax base. The scenery alone makes the trip worthwhile—and I should know, having driven every mile of it along the beach. But before you leave town, check out the piers—Dania, Lauderdale-by-the-Sea, Pompano Beach, and Deerfield. Things change for the better at Boca Raton, despite a level of development that rivals Broward County. However, Boca boasts a number of public parks—some on the beach—plus, anglers can access Boca Inlet.

Miami-Dade County

Unless you're a resident or an insider, Miami-Dade and Broward Counties are a surf caster's wasteland. I even heard a rumor that the residents of oceanfront condos are discouraged from fishing in their own backyards. Not only is fishing marginal at best, but just getting to a fishable beach is a nightmare. It's hardly worth the effort when a better opportunity lies just 30 miles north, in Palm Beach, Martin, and St. Lucie Counties.

A lone surf fisherman greets the dawn. Long rods, all in sand spikes, indicate a pompano fisherman. MIKE CONNER

On most Gold Coast beaches, daytime fishing is almost universally prohibited. However, some municipalities allow it after the lifeguards quit at 5 p.m. and before they arrive at 9 a.m.

The problem in Miami-Dade County is that the beach is obscured by a concrete seawall of hotels and condos that stretches for miles. So I forget trying to reach the surf.

Despite the fertilizer, pesticides, and who knows what else that flows into the ocean from Port Everglades Inlet, not to mention the fresh water, a few Broward beaches are worth your time—Dania and Hollywood for snook and pompano, as well as Harbor Beach.

At Hollywood and Dania, occasional runs of pompano start in late January and continue through mid April if the southeast wind cooperates (12 to 18 knots). Plus, Dania has snook during the summer. There's metered parking, if you can find a space.

If you head north into Broward, conditions start improving. Bypass Port Everglades Inlet and go north on beach road A-1-A. There's a beach, for example, just north of the Port Everglades jetties where you can still catch pompano during the fall and winter. You'll find limited access in this beachfront neighborhood, but that hasn't deterred the few anglers who are determined to fish there. As far as public parking on South Ocean Drive, the dozen or so spaces that remain are about to be fitted with meters. Get there by taking SE 17th Street east and making a right at the first light after you cross the causeway followed by a left on Mayan Drive. You face a several-hundred-yard hike before your toes touch the sand. Still feeling game? Then follow Anchor Drive to the gated path that runs just north of the Breakwater Towers.

There's a move afoot to close this beach to outsiders, but my sources in the know say that's not likely to happen. As for Fort Lauderdale's famous public beach, fishing is prohibited during beachgoing hours, but it's permitted at night. Read more about it below.

While it's difficult to enforce walk-in regulations—especially below the high-water mark—just getting to the water, unless you own or rent near the beach, is next to impossible. It's like that for miles, except where fishing from the surf is outlawed. In Lauderdale-by-the-Sea cops hand out tickets. You can thank the developers who strangled our beachfront, along with nearsighted public officials.

The situation improves but not by much, compared to what you'll find from Hollywood south. You can fish Hollywood Beach between 5:00 pm and 9:00 am, when the lifeguards are gone. Then there's John U. Lloyd Beach State Park (see page 242), where daytime fishing, however futile, is allowed in certain locations.

In the city of Fort Lauderdale, according to statute 7.4(b):

> Fishing or netting of fish is limited to the hours of 6:00 p.m. until 8:00 a.m. when on the beach and must be conducted in a safe manner. All debris, bait, fish line and hooks, and other fishing equipment or tackle must be removed from the beach after fishing has been conducted.

Beach Safety

Undertow can be dangerous to swimmers; in fact, every year undertow claims numerous lives. If you find yourself trapped, don't try to fight it. Instead, let it carry you offshore until the current lessens. Then swim parallel to the beach before heading back to shore.

Also, learn to avoid dangerous rip currents and Portuguese Man O' War—those colorful Physalia that resemble pink balloons and whose tentacles pack a powerful wallop. Along with water, carry sufficient sunscreen. And don't forget a hat.

Terry Reynard, Fort Lauderdale's former assistant director of parks and recreation, who provided this information, added:

> These rules are for all Fort Lauderdale beaches, which run from the old Yankee Clipper [Hotel] to the southern boundary of the Bonnet House property (just south of Sunrise Boulevard), then from the northern boundary of Bonnet House to Oakland Park Boulevard. The Point of Americas condo, Galt Ocean Mile and the Bonnet House are all within the city limits, but they're not city beaches. We maintain the Bonnet House, but they have an agreement to control this area.

Palm Beach, Martin, and St. Lucie Counties

Just as you enter Palm Beach County, the fishing improves. Everything changes and becomes more relaxed, starting at Boca Raton. Take Yamato Road east from I-95, all the way to where it meets A-1-A. Head a few miles north and you'll find partially submerged rock piles along with an inshore reef. Arguably the most famous is known as Jap Rock. Back when I was a kid, my friends who had driver's licenses would cast plugs there to the monster jacks that chased Spanish mackerel every year when spring rolled around. Forage still collects there. Tom Greene, who grew up in nearby Boca Raton, says snook are still available on Delray Beach during the summer, including for fly fishermen.

The city of Boca has numerous parks with some either bordering or near the beach. Parking isn't cheap there at $16 a day, although at least one facility charges by the hour. You can also park at the foot of the Spanish River Boulevard Bridge for free and walk to the beach.

Head north out of Boca, and you'll pass through Highland Beach, Gulfstream, and Delray Beach. Look around, and you'll find plenty of access if you're willing to work around swimmers and lifeguards. Walk-in fishermen can easily access beaches at Boynton, Lantana, and Phipps Park (north of Lake Worth Pier). Read the signs (although I couldn't find any) or ask a lifeguard.

Surf fishing is legal on Palm Beach Island, as long as you're not trespassing or on a public beach when the lifeguards are present. Here's what Deb Rybovich, daughter of legendary boat designer Tommy Rybovich, had to say about her bailiwick:

The beach, looking north toward the inlet, on Palm Beach Island. Don't attempt to park here, except in designated spaces.
STEVE KANTNER

Bridge to the beach at John D. MacArthur Park. STEVE KANTNER

There's a public beach in the town of Palm Beach proper that you can get to by taking Royal Palm Way east to where it dead-ends at A-1-A. (Take Okeechobee Road east from I-95, and it eventually turns into Royal Palm.) The beach is accessible for a several-block area that stretches both north and south. Worth Avenue is just to the south. Parking is available, but it's both expensive and limited. Then A-1-A ends a few blocks north where the private estates begin. While the beach road per se no longer exists, you can get to the beach by following North County Road. Take County Road north, and you'll find limited access, but you may have to park on a side street. Keep in mind, however, that parking is almost universally prohibited, and that the cops waste no time in handing out tickets. Meanwhile, a few trails or tunnels lead to secluded beaches.

I asked Deb about reaching the once-famous south jetty that frames the entrance to Palm Beach Inlet: "The biggest problem," she said, "is parking. Apparently, the locals don't want their view spoiled, so they solved the problem by doing away with motorized vehicles. You can still take a bus (Number 41) or bike to the inlet, as well as to other points on the island."

I followed Deb's suggestion and retraced her route. If you do the same, you'll encounter numerous bicyclists who come to the island to sightsee.

Take A-1-A north from Palm Beach Inlet. Go back on Flagler to U.S. 1 and take it north to Blue Heron Boulevard. Then head east on Blue Heron until you're back on the beach road. You'll find a surplus of beachfront access. When it comes to fishing, stay away from the bathers. The southernmost reaches of this barrier island, just past the rows of tall condos, is referred to as Singer Island. It's the home of several well-known beaches: Singer Island Public Beach, Ocean Reef Park, and John D. MacArthur State Park. All are worth fishing if the surf's not too rough. (If it is, head to Palm Beach Inlet. See page 244).

The beach at Singer Island is protected by lifeguards, so fishing is prohibited whenever they're on duty but only in the guarded area. Walk to the north or south, or go early or late.

Ocean Reef Park is also guarded, but you can start fishing north of the lifeguard stand. This park, which has showers and restrooms, takes its name from the rocks on its northern periphery. Fish up by the rocks where the water's deeper—run fish collect there during fall and winter.

The entire stretch, from Palm Beach to Jupiter Inlet, is a mecca during spring for schools of spawning spinner sharks.

The southern boundary of John D. MacArthur State Park lies just a few miles north of Ocean Reef. MacArthur boasts a shallow lagoon, a nature center, and kayak rental facilities. A tram runs across the $2/3$-mile-long footbridge to the beach. The beach is unguarded and fishing is permitted. You'll find restrooms and a drinking fountain at both the nature center and the beach, where a pavilion overlooks the blue-green water.

Head 2 miles farther on A-1-A, past Lost Tree Village and the Seminole Golf Course (neither of which have public access), and you enter the town of Juno Beach. You can't miss the lake on the right, and Kagan Park in the lake's southwestern corner. The park closes at dark and parking is limited. But if you walk toward the beach, you'll see the pathway that runs between the condominiums. Don't worry; it's a public beach.

The surf to the south is a good place to fish. While there's no reef per se, lime-stone formations contribute to sandbars, which control the depth and help regulate the current. You can't miss the tree to the south, and nearby you'll find a deep pocket, although the fishing is generally good between the first and second sets of rocks, too, for pompano as well as mackerel and bluefish.

Two major rock formations between MacArthur Park and Lost Tree Village disguise deepwater pockets that frequently hold fish. This entire area is bluefish country, and Spanish mackerel tend to hole up here, too. If Singer Island gets the nod for pompano and spinner sharks, then this is the place for bluefish and jacks. Spanish run the surf during the early morning, while blues become more active closer to dusk. Although mackerel are associated with deeper water, they frequently feed in the curl of the waves.

At the north end of the lake on Mercury Way, you'll find a crossover where the original Juno Pier was. We locals call it pierless, since the remnants of the old pier were eventually removed as a navigational hazard. Keep heading north, and you'll see the new pier, along with a parking lot capable of holding hundreds of cars, a well-maintained restroom, and a lifeguard office.

The bluefish are running at Coral Cove Park. STEVE KANTNER

Stay on the road and you'll find parallel parking, as well as crossovers that lead to the surf. The last time I checked, I counted dozens of access points. While fishing is legal, steer clear of the bathers.

One beach that's popular with pompano fishermen lies just across from the green-roofed condo. When the pompano are running, you'll see scores of rods lining the surf. Keep heading north to Carlin Park and another beach protected by lifeguards. Fishing is permitted outside the warning flags. I've enjoyed some stellar successes at Carlin.

Stay on the beach road, and you'll come to the entrance of the Jupiter Beach Resort. Go a bit farther to the twin entrances to DuBois Park and the Jupiter jetty. Keep following the beach road as it winds around, and eventually you'll cross Jupiter Inlet. There's lots of good fishing north of the inlet, but don't try to find it by dangling a sinker.

Coral Cove Park provides good angler access, although the bottom here is a latticework of worm rock. Anglers who fish bait at Coral Cove cast from a rocky promontory that juts from its southernmost boundary. The entire area has one thing in common: While shallow-running lures may be practical to fish here, the only sinkers with a reasonable chance of recovery are those with prongs that straighten under angler pressure, known locally as "spiders" or "Sputniks."

You can also fish the Intracoastal across A-1-A, where the water is clear regardless of the tide. The stretch between channel markers 36 and 46 (south of Bridge Road) has long been famous for yielding pompano. Bluefish and ladyfish also make a showing here.

Head north for another mile or two, and you'll come to Blowing Rocks Preserve. This area was once famous for snook, before beach renourishment—the pumping of sand from various outside sources—and illegal netting drove most linesiders from the surf. Critical bottom structure was covered by sand. However, during a recent visit, I was pleased to see that quite a few snook had returned. Nature has a way of healing itself.

Blowing Rocks takes its name from rocky outcroppings: The most prominent example is the Anastasia Formation, a coquina shell deposit stretching intermittently between St. Augustine and Palm Beach. When swells hit the rocks, they explode—leaving everything soaked with spray. It's deeper here than at Coral Cove; plus, it's possible to find patches of sandy bottom if you walk north far enough. Bluefish, jacks, and spinner sharks prefer the rocky areas, while pompano stick to the sand. There are, however, no hard-and-fast rules here. A major drawback is that the Nature Conservancy, which owns the property and charges a nominal fee for admission, locks the gate at 4:30 p.m.—just when the fish are starting to bite.

The next public access, located at the intersection of A-1-A and Bridge Road, is the Hobe Sound Pavilion. Although this incongruous structure looks a bit out of place, it's a gathering spot for anglers and beachgoers alike.

Hobe Sound is one location you can trust for action. It's where gamefish show both early and late—before they show in surrounding venues. You'll find banner runs of bluefish, along with pompano, mackerel, and the occasional deepwater species. Plus, the mullet schools stop here on their annual migration, drawing larger fish. The signature feature of this stretch of beach is an inshore reef that lies just

Fly Fishing for Snook in the Surf

Snook are an important fly-rod quarry, and by mid-July it's not uncommon to see large schools slipping down the surf line on out-of-the-way beaches, as well as those closer to home. Snook continue to patrol the surf—along Hobe Sound Beach, on Jupiter Island, or farther south at Juno Beach—well into August. You can also find schools in southern Palm Beach and Broward Counties if you're willing to walk.

Tackle includes plain bucktail or synthetic streamers such as Lefty's Deceiver tied in white, fished on slow-sinking or intermediate lines. Fly fishers who work the surf rely on 9 or 10-weight rods. See "Saltwater Fly Gear" back on page 162.

Local fisherman Larry Dringus casts plain bucktail or synthetic streamers—such as Deceivers or Clouser Minnows tied in white—and also occasionally relies on poppers. He refuses to fish with a floating line. Instead, he carries heads made from Rainy Foam that allow him to customize his regular streamers within a matter of seconds. He ties his modified lures to a 6-foot leader that he rigs on his Wet Cel II. According to Dringus, the sinking fly line pulls the popper down the face of a wave in the manner of a fleeing mullet. This approach, he says, "works well during the bait run."

Dringus also relies on tiny rattles that he inserts in a plastic tube that he adds to his

flies—but only if the surf is cloudy or roiled. He casts "either diagonal to the beach, or straight out, depending on the tide and sea conditions."

"When the surf is calm and the tide is up," he says, "you can actually see the fish with their noses practically touching the sand." Those fish are looking for food. In the surf that includes everything from tiny anchovies (or red minnows) to foot-long croakers and whiting. It's no surprise that the best snook fishing takes place during the closed season.

"August is traditionally our best month," Larry explains. "The bad news is that as the season opener approaches, strangers show up and start trashing the beach." That's when Dringus hops in his car: "The best fishing definitely takes place farther north, between Juno and Stuart, but I fish here in Broward a lot because it's where we live."

I asked Larry about overcrowding: "Not to worry— it's a big beach."

The miles of surf and seawalls, not to mention wadable flats, provide myriad opportunities for fly fishers. According to Dania Beach resident Larry Dringus, the best tide on the beach extends from three hours into the incoming through the first two hours of the ebb. That's when the linesiders work closer to the beach.
STEVE KANTNER

An angler standing on Hobe Sound Public Beach watches a hang glider pass while he casts for bluefish. STEVE KANTNER

beneath the surface—so near that, at low tide, it's barely waist-deep. The reef breaks up the groundswells, making beach fishing a lot more reliable.

You'll find the entrance to the Hobe Sound National Wildlife Preserve a few miles north at the end of the one-lane beach road. Like Blowing Rocks, it has a fee, which helps cut down on the competition. Although pets aren't welcome on this particular beach, surf fishing most definitely is. While the water isn't as deep as at Hobe Sound proper, it's an ideal venue for snook. Then, if you walk far enough north, you'll find deeper water. It's a several-mile hike to St. Lucie Inlet.

For more information on Jupiter Island, visit www.jupiterisland.com/map.html.

Stuart

To get to Stuart—and bypass the inlet—go across the Intracoastal on Bridge Road and head west to U.S. 1. Take a right at the light and head north. Then, either follow the roundabout into Port Salerno or head a few miles farther north to the Jensen

Causeway before crossing it to the beach. It's a bit confusing without local knowledge, so study a map before taking the plunge.

The beach between St. Lucie and Fort Pierce Inlets features multiple access sites. The various beaches don't differ as much as ones farther south. On Hutchinson Island, it's all about baitfish and water color. And, of course, wherever the fish are running is the best beach. Like everywhere else, fish move with the current, which is typically determined by the wind direction. Remember, wind moves water via the longshore current.

Just north of St. Lucie Inlet lies Bathtub Beach—so named because household appliances were dumped there in the past in order to stem beach erosion. Get to Bathtub (and the historical House of Refuge) by taking an immediate right after crossing the Stuart Causeway and following the road south through the golf course. Among the most popular beaches in Martin and St. Lucie Counties—for Spanish mackerel, bluefish, pompano, and whiting—are Jensen Public Beach, Walton Rocks Beach, Blind Creek, and Normandy Beaches. This entire area is part of Hutchinson Island. The Hutchinson surf is rife with sandbars, so follow the tide or arrive early or late. The closer you get to any inlet, the greater your chances of encountering runoff. The south jetty at Fort Pierce Inlet helps divert seaweed, which is always helpful when you're fishing the beach.

Here again, bluefish and pompano top the hit parade, with whiting and Spanish mackerel vying for second place. Expect tarpon during summer, along with snook (for releasing)—especially a short ways north of St. Lucie Inlet.

Fishing Piers

Few shore-based venues provide so much enjoyment and opportunity for such a nominal investment as piers. For a few bucks a head (on most piers, anyway), anglers can wet their lines in up to 20 feet of water, depending on their casting skills, without leaving the safety of a land-based structure, the pier.

You can fish for more species in the deeper water, although some require special tackle. Piers cater to three kinds of anglers: subsistence fishermen, with their buckets and dead bait; run fishermen who are after mackerel and certain large pelagics; and a third group that targets huge sharks.

The sunrise is spectacular at Dania Beach Pier. Pier fishermen—who get more for their buck than most shore-bound anglers—encounter surf-run species, plus offshore fare. King mackerel, bonito, and occasionally, a real eye-popper—including blackfin tuna, dolphin (dorado), and sailfish—have been caught on Lake Worth and Juno Beach piers and, to a lesser extent, on the Deerfield International Pier. PAT FORD

Tackle used on piers runs from bait rods to gear large enough to tackle monsters. Note the rope gaff. STEVE KANTNER

Piers along Florida's Gold Coast fall into two basic types: true oceanic piers and those with more of an inshore character. The latter suffer less from the effects of groundswells and host tropical reef life, while true ocean piers—those that lie closer to the Gulf Stream—are victimized in winter by the unobstructed pounding

The farther south you go, the smaller the swells, because the Great Bahama Bank helps block these waves. This can work for or against you: Due to the persistent wave action, the beach is packed harder at, say, Lake Worth and Juno than in Lauderdale-by-the-Sea, Pompano, or Dania. Since the constant pounding helps vet the sand grains, it takes a lot more wave action to create turbidity. That's one reason why the water stays clearer in Palm Beach County. Another is the proximity of the Gulf Stream. Due to the lack of "local color," seasonal migrants like mackerel, bluefish, and pompano don't stick around like they do farther south in Broward. (Miami-Dade does not have a pier.)

On the other hand, true oceanic predators—king mackerel, bonito, and a host of others—feel more at home in the clearer water. Plus, they'll come inshore over a white sand bottom, which you're more likely to find north of Hillsboro Inlet. But for every king that's landed on Anglin's or Dania, thousands are decked on Lake Worth and Juno. The Deerfield International Pier, the most central pier, serves up a mixture of fish.

Pier fishermen complain unceasingly about sea conditions that are too rough or too calm—while the "perfect" water lies just offshore. The reality, however, is more reassuring: After every spate of groundswell activity—when the water is choked with sand—comes a corresponding period of piscine abundance when predators from sharks to itinerant school fish flood the shallows to migrate or feed. Once the current slows down and the water clears, the fish disappear until another storm threatens.

Pier Species

Many pier anglers target food they can eat—from croakers, whiting, sand perch, blue runners, sheepshead, and grunts to such lofty fare as pompano and snappers.

Others target pelagics with live bait or lures. Their list of quarries spans the spectrum from bluefish and Spanish mackerel to pompano and permit. It also includes cobia, king mackerel, and tarpon, along with bonito and other offshore fare. Several world-record permit were caught on Lake Worth. The majority are caught in one of two ways: either on sand fleas while fishing for pompano or on calico sand crabs that were snared in the surf—in crab traps or in tangles of mono that are baited with carcasses or chunks of dead fish. Here's another tip: The majority of permit are caught at night, specifically on nights when the current reverses and runs north to south. Thirty-pound gear and a sliding sinker is standard.

Mutton snappers make appearances on piers—typically after a fall Nor'easter. The most important suggestion I can give to a novice is to watch other anglers before making a move.
STEVE KANTNER

A third category of pier fishermen, which lately has fallen out of favor, is the group that baits giant sharks, rays, Goliath grouper, and anything else that requires big-game tackle and hawser line to subdue. Shark fishing on piers is almost universally prohibited. Yet mandated closures have done little to curb it since both sharks and unrepentant sharkers still frequent these haunts. Also, you can accidentally hook one. A shark is the ultimate predator, so anyone fishing anything larger than a sand flea is inviting a strike from these seagoing garbage trucks.

You don't see as many small sharks as in the past, which suggests a nonsustainable resource. Hookups aren't as common either as they once were—with the possible exception of spinner sharks, which appear to be fighting the curve. Spinners will attack a variety of baits and lures. They are as feisty as other, betterknown gamefish.

Spinner sharks are more commonly hooked by accident than larger species are. Most spinners are hooked on sporting tackle—a solid defense for the occasional enthusiast. Whereas a 12/0 reel is a sealed indictment, a live blue runner or a chunk of bluefish sails under the radar of most pier detectives. Recent improvements in tackle such as super-braid lines, have made landing spinners a reasonable prospect—regardless of the angler's intention.

Part of the magic of fishing a pier is that you never really know what's out there. One day while fishing the Lake Worth Municipal Pier I was leaning on the railing in the shade of the shelter, looking south at bonito that were harassing a bait school. Every so often, the school would tighten and make a halfhearted romp for the pier. Occasionally they came all the way—then I cast my plug before finally losing interest.

You could see the bonito surfing down the waves, forming a horseshoe in the bait school as they strafed the sardines. Then, all of a sudden out of nowhere, a 10-foot blue marlin burst out of the swell and tore through the school. The marlin tossed a bonito at least 20 feet; then as soon as it landed inside the pier tee, less than 300 yards from the beach, she was there to grab it, amid an incredible cascade of spray.

South Florida's Piers

Some of the piers I fished are no longer standing. One in particular was the Sunny Isles (Newport) Pier, which was located at the foot of 163rd Street in North Miami Beach until it succumbed to Hurricane Wilma. For nearly 50 years, it maintained an unequalled reputation for Spanish mackerel. The mayor recently unveiled plans to rebuild it.

Anglin's Pier, Lauderdale-by-the-Sea

Melvin Anglin, who relocated his family here from southern Indiana in hopes of cashing in on the real estate boom purchased acreage that would later become Lauderdale-by-the-Sea. The pier, which was built to attract tourists and investors, proved to be a winner for both bankers and anglers. Anglin's Pier—extending from Commercial Boulevard—was erected by the Anglin family in the early 1940s. Today's rebuilt structure is at the same location.

Dania Beach Pier at dawn. Piers provide bait and sundry items, including food and drinks, as well as places to fish. Dania, Anglin's, Deerfield, and Lake Worth also have restaurants. PAT FORD

The bait shop, allegedly, opened before the pier was built. According to one account, Tom Anglin Sr. was selling mullet to a commercial fisherman when a friendly argument about his choice of locations ensued. Anglin—so it's said—was so confident in his decision that he invited Barney Skejdahl to make a cast from the deck. At the time, the boards barely reached the breakers, but Skejdahl obliged, and the rest is history.

Supposedly Skejdahl hooked a tarpon on his very first cast and managed to land the 100-pounder by dragging it through the surf. His bait? What else but a mullet fillet? Talk about instant PR!

Anglin's may be known for its tarpon, but its crowning glory lies in a pompano run that peaks during March and April. During spring vacation in my college days—when I knew no internal clock—I'd fish Anglin's for pomps in the dead of night and then catch a few winks after selling my catch, before starting all over.

In the early years, Anglin's extended out over living coral that attracted snappers, groupers, and other reef species (triggerfish and sea bass were local favorites), including a surplus of huge barracudas. However, Anglin's mainstay has always been run fish.

That fishing took place (and typically still does) just inside where the reef once existed—before pollution destroyed it. The channel that is left funnels school fish during their migrations.

Imagine you're back in the 1960s. Businessmen in shirtsleeves lined the railing, replete with nail aprons and terrycloth rags, before rushing off to work. People fished to make ends meet, just as they do today. The management iced your catch in a soft-drink cooler that they tucked away in the shop. Then when you returned after work, you retrieved your stringer and went back to fishing. Get caught selling fish—any significant amount of it—and you were summarily barred from the pier.

Feathering with jigs was popular with adults, while younger anglers chased sharks and tarpon as a rite of passage. I still the smell freshly brewed coffee, as the aroma wafts from the old wooden shop.

From the very beginning, Anglin's attracted a crowd of regulars, including many prominent, upstanding citizens: Doc Gillingham, who fished there until the age of 90; Ray Smith; Dick Hess; Claire Miner; "Mackerel Annie" Turse; and many others. These were the stars of my youth.

While a number of storms have damaged Anglin's, the management always rebuilds. Now the pier stands on concrete pilings, just like the others up and down the coast. Yet the environmental factors that killed off the reef may not change. I blame the freshwater releases that belch from Hillsboro Inlet, along with persistent attempts to renourish our beaches.

The pompano run is still worth fishing. Plus, tarpon pass by the pier from spring through fall—although they're now reluctant to hit. I landed a 151-pounder on Anglin's once on a 3/0 reel and a mackerel head. I'll never kill another.

Anglin's is the priciest pier on the coast—with a $7.50 admission fee and a $10 parking charge.

Pompano Beach Pier

Pompano Pier—located north of Anglin's, roughly where Atlantic Boulevard intersects the beach road—opened while I was a teenager. I fished there on rare occasions since the species it offered were similar to Anglin's, albeit with a twist or two.

Seasonal migrants, which I enjoyed on Anglin's, were less common at Pompano, but huge snook swam through the lights at night, while tarpon fresh from Anglin's tooled north past the tee. I won free admissions on several occasions in monthly contests and caught a 70-pound tarpon, a 35-pound cobia, and finally, a 31-pound snook. The deck, I admit, was stacked in my favor, since I never saw anyone pursuing these fish. I'd bet that's changed by now.

I also saw kingfish—including one that my friend caught—which were in short supply at Anglin's. But the best news of all, for a college student like me, was an arcade near shore that housed a pub. Unfortunately, the pier's been rebuilt and the pub no longer exists.

One thing that makes pier fishing different from everything else in this book is that you don't need a fishing license. Piers—like charter boats—purchase blanket permits that cover anyone who fishes from their decks. So someone visiting from Montreal or Topeka can fish these venues without buying a license. It's one way our

A hole in the railing provides access for handicapped fishers. STEVE KANTNER

state encourages tourism, and it's a great idea. Who, after all, thinks about purchasing licenses when they're having fun—and, in the process, contributing to the state's economy?

Pier admission is free, and metered parking across the street only costs a buck an hour. But there's better news from a fishing standpoint: I fished Pompano several mornings this fall (it's cheaper than Anglin's) and came away with a $5\frac{1}{2}$-pound Spanish mackerel, along with several others, after releasing small pompano. I saw sharks and porpoises and acres of ballyhoo.

Deerfield Beach International Pier

Deerfield extends seaward from A-1-A, just north of where Hillsboro Boulevard crosses the Intracoastal. I first walked the boards on the old pier when I was 11 or 12. A sign from that initial visit shaped my future fishing efforts. It read: "For better fishing, try spinning."

Now there's an idea. I'd been struggling with bait casters and braided nylon or ready-made leaders and salty shrimp tails. My triumphs were measured in sand perch and puffers ("toadies" in Palm Beach County parlance) when a fisherman on Deerfield landed a 20-pound kingfish. He was fishing live bait on spinning tackle.

My imagination was spinning as fast as his reel spool. Within a matter of months I owned a reel like his (a small Orvis knockoff known as a Regis), a 7-foot rod, and a box full of lures. The world around me had started to change. By the time I graduated from junior high school, I was hitching rides to Deerfield whenever I could. Deerfield differed from Anglin's in many respects.

I remember kingfish swarming the pier at dusk, snapping up jigs like a school of piranhas. By my junior year, I was hooked on those kings during the winter while I waited for the tarpon to return to Anglin's. When, exactly, did they return? You could mark your calendar for Memorial Day weekend. In the years that followed, I'd land at least 20 kings per season, including a 31-pounder that hit a Hopkins spoon. Then the draft board sent me a notice.

Looking back, I recall shoals of bluefish and other wonders—some, admittedly, that are hard to believe. In the old days, the management maintained its own shark rod that was available to patrons for the asking. Fishing, after all, was all about fun. Try asking for a shark rod now on a pier.

Initially, the pier was around 500 feet long, placing it over some fairly deep water; hence all those kings and some monster sharks. But a storm churned out of

Deerfield Pier in its newest incarnation. STEVE KANTNER

the Caribbean and left the Deerfield Stub, maybe a third the length of the original pier. Deerfield languished in the years that followed; however, a full-length replacement was eventually built in the late 1960s.

I was a regular there during my college years, since it was closer to home than either Lake Worth or Juno. Plus, Deerfield had fewer patrons, so with less competition, my odds improved.

The most kings (big ones) I landed in a single day was three, including the one on the Hopkins spoon. However, I watched Mrs. Grace Harvey reel a king to the pier that literally broke the gaff when someone attempted to lift it over the railing. It weighed 70 pounds if it weighed an ounce. While Deerfield wasn't the stuff of legends, it provided enough action to keep me busy.

I caught kings there on lures, as well as goggle-eyes, but the majority hit live ballyhoo that I fished on the bottom or dead ones that I drifted and jigged. I didn't realize it at the time, but I'd discovered snobbling (see page 213). There were snook and bonito, and I lost a dolphin.

In January of 2012, another angler was free-lining live bait and landed a 24-pound dolphin off Deerfield. Steve Waters, of the South Florida *Sun-Sentinel*, ran the photo in one of his columns.

On another occasion, during a northwest gale, I rigged three round balloons together to carry out my bait on a strong northwest wind. I ended up with a 25-pound cobia. I've used multiple reels on other occasions, and once I landed a whopping a 6-pound king off Lake Worth—maximum effort for minimal reward.

Other species that I fished for on Deerfield Pier included sharks (and inadvertently, stingrays). These giants frequently trailed schools of cobia. One of my favorite pastimes was to balloon out shark baits on those winter nights when Floridians shiver. Without sufficient protection, a person can die in 40-degree temperatures and a 30-knot wind.

The night man held forth in the shop, while my friends and I would balloon out shark baits. Then, at the first indication we had hooked a stingray or a dusky shark—both which were known to travel with cobia—he'd lock up the gate like the Great Wall of China and toss boxes of squid in the shrimp tank. Within minutes the squids were thawed and ready.

Once we reeled the shark or ray to the pier and saw signs of cobia, we'd go to work. We'd dangle whole squids on heavy tackle, while taking turns holding the shark rod, which by now was braced on the railing. When a cobia hit, we'd hold its head out until someone buried the gaff. We suffered multiple mishaps, which included losing the gaff. But when the action was over, we'd clean up our mess and release the shark by unsnapping the swivel. A short time later, we had fillets on ice.

The next day we'd head for the charter docks with the fillets washed and bagged. The real "hook" came when our "personal representative" purchased a set of white deck hand's attire—adding to his nautical charm. He'd arrive just in time to intercept customers who were waiting to buy cobia from the returning charter fleet. Then, just as the boats were rounding the jetties, he'd complete his transactions and hop in his car. He even had the temerity to borrow their scales.

Timing was critical, but he'd pull it off with scarcely a moment to spare. Then he'd split the proceeds with whoever participated, and we'd all buy beers at the lounge that was conveniently located across from the pier. Talk about a sustainable fishery!

I'm forever captivated by nautical novelties. At the top of my list are those blue-water wanderers that occasionally show up in the shallows. While Lake Worth Pier must hold some sort of record, the following story puts Deerfield in the running.

Young Ken Wagner—confined to a wheelchair—decked a sailfish without assistance, except for the gaff. The sail, I recall, weighed 50-some pounds; I arrived shortly after it was carted away. Ken was free-swimming a goggle-eye that he bugged near shore before wheeling it out to the tee. How did he hook that fish? The same way he did with other gamefish—by backing his wheelchair as fast as he could.

Lake Worth Municipal Pier (Lockhart Memorial Pier)

The city of Lake Worth opened this beauty in 1959 or 1960, adjacent to the Lake Worth Casino. Take the Lantana Bridge across the Intracoastal and head north on A-1-A. Of all the piers I've fished in my life, none taps into so massive a resource.

Lake Worth Pier (Lockhart Memorial Pier). STEVE KANTNER

From whale sharks to sailfish, dolphin to kings, if it swims in our waters, I've seen it here.

I've made numerous references to the diversity of species here. It's due to the proximity of the Gulf Stream, along with the pier's ability to attract large schools of baitfish. The hard sand bottom also plays a part, since the water stays clear when groundswells kick up.

In the early years, live bait meant goggle-eyes that gathered in the shadows in thick, black schools. You'd catch them before sunrise on a string of quills, and keep them alive in cages or baskets. Once the sun was up, you'd need a goggle-eye bug, the best of which was perfected by a local realtor named Ernie Viers and was known as the Pink Lady.

In the mid-1960s, schools of whitebait arrived, bringing with them a surge in kingfish. On a fateful day that I spent with my girlfriend, anglers on Lake Worth decked a staggering 269 kings. How can I remember that number? Guess. The girl dumped me and ran off with my friend. But I'd love to see another one like her— that day, I mean.

Back in the 1960s, a group of locals got together and formed the Lake Worth Pier Big Game Club. Shark-fishing contests pitted their members against the Palm Beach Club. Contests were typically held on neutral territory (for example, Juno Pier).

I've hooked sailfish on plugs, landed tuna on live bait, and narrowly missed seeing a pair of giant bluefins. Both grabbed live blue runners before vaporizing twin 4/0s. In the words of one onlooker, "The spools are still spinning."

A 36-pound bonito was landed there, and a broadbill swordfish was hooked and lost. And more big jacks and cobia than you could ever count, along with the former all-tackle world-record permit—a 52-pounder.

The greatest indication that the fish are still there is the city's willingness to keep rebuilding the pier whenever it's damaged by storms. A photo that hung in the shop at one time showed a swell about to break over the tee. The crest was 6 feet above the railing—the largest I've ever seen, not counting on the Discovery Channel.

New Juno Beach Pier

Juno is best known for snook and bonito, but it's gaining a reputation for both king-fish and tarpon. In fact, sailfish, dorado, and even tuna (blackfin) have been landed here. Maggie Luneke was the first to subdue a sail here—believe it or not, on live bait fished on a sinker rig. Juno, like Lake Worth, is at the mercy of groundswells. But because it is farther north, it gets the fall runs first. Then, once the surf calms down in spring, Juno is quick to attract schools of whitebait and predators that follow hot on their heels.

As I suggested, Juno is also a winter pier, with strong runs of bluefish, Spanish mackerel, and pompano. And it still contains plenty of croakers and whiting, blue runners, sand perch, and snappers. Juno caters to a crowd of subsistence fishermen, along with big-game enthusiasts.

Parking in the lot across the street is free. To get there, take Donald Ross Road to U.S. 1 and head north for approximately one mile.

Built after years of budget haggling, this county-owned pier replaces a tiny structure that sat a few miles to the south. Both piers played a role in providing anglers with first-class fishing for a minimal investment. STEVE KANTNER

Dania Beach Pier

Like Lake Worth, Dania was erected by a municipality alongside a public beach. I vaguely remember the opening—and that the pier nearly burned down during a fireworks display. This pier has a reputation for Spanish mackerel that gather during the winter. Anglers also catch a few snook, tarpon, and pompano.

A shark-fishing friend from my youth, John Wolmer, caught a monster hammerhead from Dania Pier. The shark, which was estimated at close to 600 pounds, paled next to the one we double-teamed on Deerfield that bottomed out a balance beam scale at 650 pounds—with its pectoral fins still on the ground. We no longer kill these sharks.

To get to the pier, take Dania Beach Boulevard east across the Intracoastal and follow the signs.

Captain Jimmy Duncan on the original Lake Worth Pier, with a sail he nailed on a discarded goggle-eye. Piermaster Captain Ralph McKeral is on the right. CAPTAIN JIMMY DUNCAN

Jimmy Duncan with a typical king mackerel during the heyday of Palm Beach Pier. The pier has been gone now for several decades. CAPTAIN JIMMY DUNCAN

The old Lake Worth Pier Big Game Club. CAPTAIN JIMMY DUNCAN

Essential Pier Rigs and Methods

Most piers allow their patrons a maximum of three rods apiece. Veterans, however, use this rule to their advantage by rigging each one differently as the season progresses. This quota usually equates to a pair of spinners—a one-handed model that doubles as a bait rod and its heavier cousin for casting large lures—along with a medium- or heavy-action conventional outfit: say an 8- to 10-foot heavy-action conventional rod matched with a conventional reel capable of holding at least 300 yards of 30-pound mono.

Let's take the spinners first. Most anglers who cast jigs or spoons for run fish—mackerel, bluefish, and pompano—rely on a 6½- to 7-foot, medium-action spinning rod and a matching reel that's capable of holding approximately 250 yards of 10-pound-test mono (or its super braid equivalent). Let's refer to this outfit as spinner #1. They can use spinner #1 for catching whitebait by adding a lightweight, pre-rigged Sabiki. Most run fish, remember, are gone by summer.

More ambitious anglers who target kingfish, large jacks, and bonito prefer something stouter—both for the heavier lures and the tougher fish. This also applies to free-lining live baits.

For either, I recommend a 7½- to 8½-foot, medium- to heavy-action spinning rod matched with a high quality reel that's capable of holding at least 300 yards of 12-pound-test mono. Let's call this outfit spinner #2.

Pier fishermen select their tackle based on the quarries they intend to target; however, most piers only allow three rods, so savvy anglers must make sure their rods and reels perform multiple functions. To change lures quickly, I attach a snap swivel to my main line with a 9-inch Bimini loop. The loop protects the line against nicks and frays, while helping to maintain the line's original test. PAT FORD

Spanish mackerel like this one are pursued on piers—occasionally year-round if conditions oblige. Small jigs, plugs, and spoons are what pier fishermen rely on when targeting school fish like bluefish, pompano, and Spanish mackerel.
PAT FORD

A smooth drag is essential for spinner #2's sweat work. A 15- or 20-pound-test may be a better alternative, depending on your quarry and the size of the crowd. Then there's always super-braid. Try 30-pound PowerPro.

A longer rod may be easier to cast, but if it's too long, it gets in the way when you're fighting a fish. So what would I call a sensible compromise? An 8-footer is fine for piers, but not if the butt cap protrudes much past your elbow, once you position the butt beneath your forearm. That makes it easier to move in a crowd.

In the fall, after the pilchards, thread herrings, and Spanish sardines have all departed, fishermen use their bait rods for jigging, particularly for pompano. While the majority of the pier crowd pursues pomps with surf rods—which they cast a country mile—jiggers slip between them and their telephone poles and catch their limit on this lighter gear. What then becomes of spinner #2?

They may use it for snook with a live shrimp and sinker—or for any fish that requires such a setup. The list includes mangrove and mutton snapper, pompano, bluefish, and tarpon—the latter occasionally eat castoffs from cutting boards.

And now for outfit #3. Anglers use conventional outfits during the summer for fishing live baits with a sinker, trolley, or occasionally, free-lining. But the whitebait schools have taken off for the Keys, or wherever they go at the onset of winter—leaving an outfit behind that could use an assignment.

You could use it to target everything with from spinner sharks to cobia. Or skip the sinker and use it for fishing big baits—say, by free-lining a live bluefish, blue runner, mackerel, or mullet for monsters that lurk beyond the pier tee.

Fishing Live Baits on Piers with a Sinker

Let's say it's summer again, and I need to anchor a pilchard, sardine, or greenie (threadfin herring) on the bottom—the most common way of presenting three live baits. I attach a short piece of 60-pound mono to a 5-ounce bank sinker with an improved clinch knot before tying a black anodized barrel swivel to the opposite end. The resulting combination allows a live bait to swim up off the bottom, distancing itself from the sinker. I leave no more than 2 inches between sinker and swivel. Afterward, I slide my running line through the swivel before knotting it to my leader with a uni knot or hangman's knot.

I prefer my mono leaders to test twice the breaking strength of my running line, plus an extra 10 pounds. That way, I stay on the safe side. For toothy fish, I add a 4-inch piece of coffee-colored, single-strand wire that tests twice the strength of the running line. Here's how I rig it: First, I put a sharp bend in the wire—a full 180 degrees—before I bend the loop with my pliers and join the mono to the wire with an Albright knot.

Wire sizes are expressed numerically, and they're based on diameter rather than strength. Still, you'll soon discover soon that, regardless of the manufacturer, #2 or #3 (the smallest commercially available) tests somewhere in the neighborhood of 30 pounds, while #7—another commonly used size—tests around 70. Single-strand is less visible than some braided alternatives.

With a sinker, I limit my leaders to 14 or 15 inches, which prevents the bait and sinker from separating in the air and shortening my cast. It helps if the sinker weighs more than my bait.

Free-Lining Live Baits

Free-lining is by far the most popular method of presenting large live baits like blue runners or goggle-eyes. It's also effective for fishing whitebaits hooked through the nostrils, just ahead of the tail, or the "heart" (actually a piece of cartilage that's located directly in front of the pectoral fins), depending on where we want them to swim.

Live baits fight resistance by attempting to swim in the opposite direction. This can be used to your advantage to send them out to sea.

It takes some practice. Picture the upcurrent corner at the end of a pier. You start out by casting a tail-hooked live bait seaward as far as you can—and slightly upcurrent—before eliminating slack by reeling.

When you can feel your live bait, take your reel out of gear. You want it to take out as much line as possible while you provide resistance as a point of reference and prevent any overruns.

Keep your thumb on the spool flange (if it's a conventional reel) and not on the line. Monofilament scorches raise nasty blisters, and some predators strike live baits

Shaved thread herring will be free-lined for mackerel. PAT FORD

with the force of a train. You'll want to swim your bait as far out as possible before the current acts on both it and your line.

Back when free-lining was big on the Lake Worth Pier, everyone used special tackle. We'd spend hours before daybreak gathering goggle-eyes that we'd send off the tee at the first hint of daylight. Large-capacity reels filled with 400 or more yards of 20- to 30-pound line were our weapons of choice for everything from kings, cobia, jacks, and bonito to dolphin and sailfish—in fact, the pier earned fame for its Atlantic sailfish.

Free-lining requires skill as well as commitment. If the practice seems to be dwindling on area piers, it may be due to the increase in boat traffic and the boats driving over your line. But it's still highly effective. For the past several winters, anglers on Juno have been free-lining live bluefish. They landed some spectacular kingfish as a result. And at Anglin's, free-lining for tarpon remains a summer tradition.

Some predators, such as migrating tarpon, require lively bait—for example, a nose-hooked pilchard. Youthful anglers on Anglin's Pier, and to a lesser degree on Pompano and Dania, start by catching pilchards on Sabiki rigs, which do minimal damage, before hooking their baits through the nostrils—which they accomplish while holding the bait's mouth shut. I've fished in this manner since I was 13, so I've seen techniques evolve.

Say it's the first week of June, the southeaster's blowing, and whitecaps are rippling the surface while a strong north-running current lifts a few dangling bait buckets. Your watch says it's four o'clock, and beachgoers are leaving; it's like every other day in Lauderdale-by-the-Sea as evening approaches. The pier, which was buzzing with multilingual chatter, is now silent except for the wind and the gentle clicking of some noisy baitfish that move back and forth from the shed to the tee.

You reach for your bait rod and jig up a fat one before you take off running to where you left your free-lining rod, a custom-made spinner from a local shop. After hooking your pilchard, you shove aside a few buddies who are already waiting in the southeast corner. Then you give your bait an underhand toss.

The air feels charged—perhaps the remnants of a recent shower. Your bait digs deeper and then suddenly is forced up on top by a 6-foot tarpon that vacuums it in while it heads north. Then it's one, two, three, and your line's on the roller as you step in to the fish with maximum force. Your rod bends precariously, and your reel starts to stutter before emitting a high-pitched whine. Now you watch your tarpon stand on its head a hundred yards from where you just hooked it.

If you'd like to try this type of fishing, visit one of these piers in the late afternoon, from the middle of May through late July. The tackle continues to improve, but the tarpon seem increasingly dour. It's definitely not a sport for beginners.

You'll need a heavy spinning outfit like the one I described. For terminal rigging, I recommend a 3-foot length of 30-pound monofilament tied directly to a foot of 100-pound mono with a double surgeon's knot. For hooks, I like a 4/0 or 5/0 Mustad 9174. It's customary to reel a beaten fish to the pilings and then grab your spool in order to break it off. Be sure you're wearing a good pair of sunglasses in case the hook comes out.

Trolley Rigs

I've noticed a trend in which fishermen use no-brainer strategies—ones that require less physical involvement. Trolley rigs are its poster children. Here's how they work: The angler casts a sinker, typically a heavy one, as far as he can; then he waits until it's buried in the bottom. Once it sticks, he clips a snap swivel over his line that's attached to a pre-rigged leader. The leader may be festooned with floats and such, but it's always baited with a nose-hooked live bait or one hooked in front of the dorsal.

When the angler raises his rod tip, the bait slides downward and works its way out toward the sinker. Eventually, it's away from the pier. What's attractive about this rig is that it enables someone with limited experience to suspend a live bait—one that hasn't been injured by casting—in a static position that's unaffected by the current. What's unattractive is that some trolley riggers lash PVC tubes to the railing, which prevents other anglers from getting around them. Also, with both the dangling bait and line to contend with, other anglers in the vicinity face the threat of tangles if they hook a fish that runs past the trolley rig.

If trolleys have a redeeming value, it's their ability to generate kingfish strikes. Those strikes, which take place at or near the surface, are truly explosive. Kings hit trolley-rigged baits so hard that they hook themselves before the line comes tight. Someone told me about fishing for tarpon this way. I can only imagine the mess.

The Sabiki is a multiple-hook rig with hooks covered with processed fish skin or possibly an artificial substitute. Colorful thread wraps help attract the bait. Sabiki rigs come in different sizes, which cover the spectrum from fully-grown goggle-eyes to tiny summer-bait pilchards.
STEVE KANTNER

Sabiki Rigs

Years ago, an angler discovered that it was possible to catch whitebait (pilchards, thread herrings, Spanish sardines, etc.) on what we used to refer to as a minnow jig. This jig consisted of three tiny hooks wrapped with colored yarn, jigged up and down in the school of baitfish. Anyone could load up in seconds.

A few years later, and quite by accident, the crowd I grew up with on Anglin's Pier discovered that this rig worked even better if we used undressed hooks. We'd slide three or four size 12 silver O'Shaughnessy hooks over several feet of 4- to 8-pound-test mono (4-pound-test has been readily available only since the 1960s). The hooks were suspended on 1-inch loops, each of them 8 or 9 inches apart. The last loop was approximately a foot above the sinker.

We discovered that if we used too few hooks, baitfish wouldn't hit, while even one too many caused the rig to tangle. Three or four hooks remained the optimum number until the introduction of rods that housed both hooks and sinker internally. These are known as Sabiki rods. Then, as now, a ⅜- to ¾- ounce dipsey or bell sinker was the weight of choice for a bait rig.

This rig continued to evolve. We tried everything from gold No. 89 Eagle Claws to blued Aberdeen long shanks, but the baitfish kept getting smarter. Having a live bait whenever we wanted one eventually became a luxury—until the advent of the Sabiki rig.

A New Look at Jigging

If you're targeting pompano instead, a yellow (or pink) ball or bullet-headed nylon jig, like the legendary Nylure, is hard to beat, as is a recent arrival known as Doc's Goofy Jig. The Goofy Jig has a curved metal body but no skirt, although some models come equipped with a dropper fly that's attached to the eye with a split ring. Like the majority of lures used on piers or in the surf, Goofy Jigs come in different

Ball Rigging

Ball rigging—a method used almost exclusively on piers—keeps a live baitfish under control while allowing it freedom of movement. It uses a rubber ball, which serves as a float and is heavy enough to cast. But instead of restraining the bait near the surface, the ball allows it to swim about freely while the line slips through a swivel. That, in turn, keeps the bait looking lively—resulting in extra strikes.

The ball rig is deadly on smoker kingfish. When these rigs were more popular pier fishermen used them in conjunction with strong-swimming baits such as thread herring and sardines, hooked through the dorsal or nose.

Ball rigs are effective, inexpensive, and easy to assemble. Here's how to put one together:

1. Purchase a 3-inch hard rubber ball in the toy section of a department store. Buy one that's easy to see in the water. You also need a 1/0 treble; a short piece of No. 12 or 15 wire; an inexpensive, medium-size snap swivel; pliers; and wire cutters.

2. Join one end of the wire to the treble with a series of barrel wraps. Don't bother making a haywire twist.

3. Be sure your wire is straight, and then trim the opposite end with wire cutters. Carefully force that end through the center of the ball by pushing it with pliers. Grip the wire close to the ball, keeping your hands out of the way.

4. Use your pliers to grip the wire and pull the treble into the rubber.

5. Slide the snap over the free end of the wire. While compressing the ball against a hard surface, secure the snap swivel in place with a few wraps of wire. Afterward, when the rubber relaxes, only the swivel portion should protrude from the ball.

6. Make your leader by attaching 3 feet of 50-pound mono (or heavier, if you're fishing with line that tests stronger than 30) to a short length of No. 6 brown wire (again, only if you're using light line). Join the wire to the mono by pinching a tight bend in the former and affixing the mono with an Albright knot. Next, attach the hook by using a haywire twist. I prefer standard short-shank O'Shaughnessy hooks, such as the Mustad 9174. As I mentioned earlier, pier anglers use this rig with a nose-hooked thread herring or greenie. With a typical greenie, a size 4/0 or 6/0 hook makes a perfect match.

The advantage of a ball rig is that it's heavy enough to cast well away from the pier, while allowing a nose-hooked live bait freedom of movement.

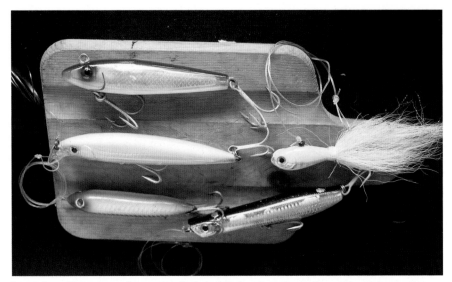

Larger lures that are popular on piers include full-size plugs, heavy jigs, and certain oversize spoons. Since these lures are heavier than those I previously described—and so is the quarry—a tackle adjustment may be in order. STEVE KANTNER

It's possible to enhance a jig's effectiveness by adding a dab of color; try red or chartreuse paint on its nose. If it's an all-white jig that you're using for mackerel, chartreuse gets the nod. Or try painting the sides of a Krocodile chartreuse, after applying a white primer undercoat. Then dip the nose in hot-orange enamel. Both colors stand out when the water's cloudy. This trick also works with powder paint. Assortment of jigs, counterclockwise from top left: heavy bucktail with surgically sharp hook, a custom version of the sidewinder, custom made pompano jig, hackle "chicken feather," lightweight homemade "crappie jig," Goofy jigs in various colors and sizes. STEVE KANTNER

sizes. Give me a ³⁄₈-ouncer on any pier, and the ⁵⁄₈-ounce version for fishing the surf, depending on the wind and sea conditions.

Snobbling with X-Raps

Adrenaline fuels this style of fishing for king mackerel, which begins when shadows first descend on the beach. The peak time is an hour before sundown, when tightly packed baitfish start crowding the tee. The schools gather in the shadows before heading offshore. Then, an instant later, they're back again—this time on the downcurrent side. You won't see movement in the vacuum behind them, just a well-defined wall of squirming sardines. Go ahead and cast. Now simply allow your X-Rap to swim in the current. Snobbling is a technique where you barely move a dead bait or the X-Rap, only occasionally turning the reel handle.

Choosing the proper retrieve is chancy with kingfish on piers, since both steady reeling and snobbling lead to trip-hammer strikes. Pier and surf veteran Terry Luneke explained snobbling to me after nine anglers nearby hooked kings in rapid

King mackerel, many larger than this one, make regular appearances on several area piers. Look for predators first on the upcurrent side. However, this doesn't apply to sharks, rays, and the cobia that follow them or the kingfish that approach by swimming into the current. STEVE KANTNER

Barracuda Tube Lure

First-timers inevitably ask this question: Why don't barracudas behave more aggressively around fishing piers? While these "tigers of the sea" attack fish once you've hooked them, they're reluctant when it comes to most live baits, as well as dead ones. So you must make them an offer they can't refuse.

First, obtain a dead needlefish that's approximately two feet in length. (I typically catch them with a bobber and strip rig). Then, bend it forcibly in order to break its backbone. When the late Mr. Needle is completely limber, cut off both his upper and lower beaks before forcing a 7/0 or 8/0 hook from the bottom to the top of his skull. Jig this bait like you would a jerk bait. When a 'cuda bites its tail off, wait. He'll come back and gobble the part with the hook.

Nearly as effective are latex tube lures—the trick is to construct one that swims like a needlefish.

First get a 1½-foot piece of green-chartreuse surgical tubing, a short length of #10 leader wire, two 4/0 reinforced treble hooks, a ¾-ounce egg sinker, and a small, black swivel. The surgical tubing is available from most tackle shops and from retail catalogs. You'll also need a knife with a sharp point and a pair of pliers.

1. Measure 14 inches of tubing before slicing away the excess with an even, sideways cut. Then measure backward 4 inches from the front and make three evenly spaced, lengthwise cuts. Each slice should be approximately 1 inch in length.
2. After cutting the tube, attach one of the trebles to an 18-inch length of wire with a haywire twist. Since you'll be pushing the wire, it's best to clip rather than kink it off.
3. Then repeat steps 1 and 2 with the second treble and a shorter, 10-inch length of wire.
4. Next, insert the longer wire into the rear of the tube (the opposite end from the three lengthwise slices) and push it until the free end protrudes from the other side. This secures the treble with the wire inside the tube.

succession—while I never got so much as a swirl. Seems I never stopped reeling to watch what they were doing (or, in this case, not doing).

You ponder why you've remained inexplicably hitless in silence. Then, while you're leaning there daydreaming, a king tries to jerk the rod from your hands. The guy standing next to you gets caught off guard, too. But you both hook fish that make sizzling runs, and you scramble to follow yours along the railing. Once you get situated, you hear shouts from behind, and you turn to see other anglers that have hooked up, too.

Meanwhile, there's more to X-Raps than snobbling. Big jacks will chomp wobblers if the plug is large enough, as will less aggressive snook and cobia. In fact, practically everything hits a swimmer if it gets the chance and if the plug resembles what they're feeding on. So don't throw them an X-Rap 14 if they're chasing small pilchards.

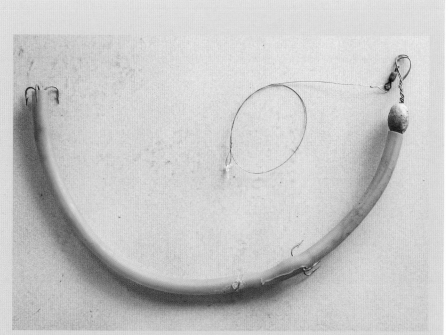

The barracuda lure, made from surgical tubing, has proven its worth on area piers.
STEVE KANTNER

5. Position the second treble by slipping the shorter wire inside one of the lengthwise cuts and sliding it forward until the free end protrudes. Using your pliers to pull on the wire, position the second treble inside the tube.
6. Slide a ¾-ounce egg sinker over the two protruding wires while snugging it against the tube. Then haywire the swivel in front. This completes the lure.

Jigs and Spoons

The problem with lures that have built-in action is their lack of weight and durability—if the lip breaks, you're out of business. When it comes to jigs and spoons, you don't have that problem.

Schools of pelagics once stretched for miles. That's when anything with a hint of flash or color got all the attention an angler could ask for. Jigs and, to a lesser degree, spoons were the tools of commercial hand-liners who earned their living bugging Spanish mackerel and kingfish. Bugs are essentially unweighted jigs.

You can only work a jig in so many ways. Fishermen on piers target migrating cobia with the same bounce-and-reel that they use for pompano. Only with cobia it's sight-fishing in its purest form, which relies on an angler's ability to spot the stingray and manta rays that serve as reference points for the pods of cobia. This type of fishing has been perfected on the piers of the Treasure Coast and Florida's Panhandle.

Snook on a pier. Remain on the alert for signs of activity: swirls, diving birds, or nervous baitfish. And while you're at it, learn to tell the difference between dappling bait (when the water's calm) and the exploding-snowball or black-horseshoe effects that signal the onslaught of feeding gamefish. PAT FORD

Jigs used for cobia weigh between 2 and 4 ounces and are rigged with a plastic worm, an imitation squid, or the mantle from a cephalopod (a neat trick). Cobia fishing is seasonal, and here in South Florida, it peaks during springtime.

A jig, it's been argued, will catch practically anything, provided the angler knows how to work it properly. Simply casting and reeling is the most basic approach; however, finding the right speed for the retrieve is always critical. If you reel a jig near the surface and big fish are feeding nearby, schooling species like jacks, bluefish, and false albacore will invariably take a whack at it, as will solitary types that you wouldn't expect, if you manage to catch them up on top. Barracudas and snook are two examples, followed by tarpon, blue runners, and certain offshore wanderers like dorado and sailfish. And the single hook nails 'em better than any treble.

A more stealthy approach—like when you're targeting deep-running tarpon or cobia that are shadowing stingrays—requires finesse. You should lead your target, rather than bumping into it. Casting jigs for snook lies somewhere in between, although it usually requires steady reeling. Gauge the fish's reactions and adjust accordingly the next time you cast.

Spoons act like jigs in several respects, but they aren't as effective in murky water. I've heard a number of stories extolling their merits, including how the founder of the L&S Bait Company Howard LeMaster, developed his original MirrOlure after watching commercial fishermen catch seatrout on spoons. Still, I'm partial to jigs over spoons—at least, in the larger sizes. And sometimes a jig will out-fish a plug.

Sometimes a nylon or hackle jig—or a spoon, for that matter—can be just as deadly as more sophisticated baits. Large spoons for using on piers, or in the surf, include the 1½- and 2½-ounce Krocodile and Gator wobblers and the 4-ounce Hopkins No-Eql. There was a time not so long ago when I used heavier Krocs, but I no longer see them in stores. Luhr-Jensen, the company that makes these lures, stamps out the bodies in various lengths. Look for the thicker versions of the 1½-ounce size.

Allow Krocs and Gators to sink momentarily before beginning a steady retrieve. If the spoon hops up on top too soon, drop them back and start over again as soon as they hit the bottom. Skip the Hopkins like a topwater plug. All of these lures can also be jigged.

A variation of the spoon is the diamond jig, which resembles a cross between a wobbler and a jig and incorporates the advantages of both. It casts like a bullet and gets plenty of strikes.

For mackerel and bluefish, you need a short wire leader, but you don't want something that will scare away fish. I start out by attaching a 2-foot section of 30-pound mono to my main line with a double surgeon's knot. The mono should be at least twice the pound-test of your main line, with another 10 pounds added for good measure. I attach a 4-inch piece of single-strand wire (no larger than No. 3,

The fishing pier beneath the Jensen Causeway—which spans the Indian River Lagoon—hosts runs of Spanish mackerel, bluefish, and pompano. MIKE CONNER

A circular net is used to land fish on piers; anglers frequently supply their own. STEVE KANTNER

marked "camouflage") to my lure with a haywire twist, before forming a haywire loop in the opposite end. Then I tie an overhand knot in the end of my mono. After passing it through the wire loop, I make a fisherman's knot with 1½ turns. I lick the mono before pulling it tight, making sure the overhand knot acts as a stopper. If I need to change lures, this one's easy to untie, and there's not enough to it to draw unwanted attention.

Landing Fish

In years past, piers provided their patrons with an effective means of landing large fish—gaffs, nets, and so on. Those days are gone now, and if piers own such devices, they're sequestered in the bait house where they're carefully guarded. The only sensible approach is to carry your own.

Some anglers construct pier nets that are mounted on hoops and joined to a rope with a float at one end. I bought a ready-made version at Custom Rod and Reel in Lighthouse Point for around 40 bucks. Once you've either bought a net or made your own, you're probably reluctant to let strangers use it—especially for a toothy barracuda or kingfish. That's why a suitable alternative is a bridge gaff.

The best bridge gaffs I've seen are the simplest. Tackle shops still sell weighted snatch hooks (despite the negative flap over snagging). The heavier models work in a pinch. I don't like the ready-made ones that slip over your line—not in the wind or during daylight hours—because they've caused many fish to break off. In the past, I bought four giant shark hooks and had a welding shop join them together.

If you've reeled up a nice fish and don't have any of these choices handy, you can try to flip it. Lean over as far as you can, take a few final cranks after tightening your drag, and then attempt to lift your fish in one smooth motion while stepping back away from the railing. If you tie good knots and your leader's intact, this usually works with a fish half your line test.

Docks and Bridges

I n many ways, docks serve the same function (for the angler and the fish) as bridges—they are simply smaller versions. Docks, like bridges, provide shelter for fish, as well a welcome respite from the relentlessness of current. They also attract bait.

When South Floridians think of fishing a dock, it's typically from the deck of a flats skiff—casting toward the dock instead of actually standing on it. That's a popular approach to light-tackle snook fishing. But there's really no reason you can't fish docks from the shore, from the dock itself, from the one next door, or by

St. Lucie sunset. Docks along the St. Lucie and Indian Rivers attract forage, which draw big gamefish. Fishing the dock lights is a popular nocturnal pastime here. PAT FORD

wading. The challenge lies in getting permission, and once you have it, not waking the neighbors.

The best docks to fish are usually near inlets because of the forage, the current, and a better shadow line. You'll find more of all three of these as you near an inlet. During the day, docks provide pleasant action for species such as sheepshead (or convict fish), an inshore species that's a popular panfish. These members of the porgy family feed on barnacles, other mollusks, and crustaceans. Use a scraper to chop barnacles from the pilings, and then gather a few to crack with a hammer before baiting the field with the shards. Every sheepshead in the area will come running. Then you can catch all you want with fiddler crabs.

You'll find good sheepshead fishing from Lake Worth north. Be sure, however, to use a small, strong hook and line that's heavy enough to wrestle these fish from the pilings.

But on docks, as on bridges, the real action turns on at night.

Under the Lights

A brightly lit dock acts like a magnet. It attracts shrimp and all types of baitfish that dance through the glow in a garish ballet. Mullet, pilchards, and Spanish sardines all join in this midnight madness that reaches its peak from May through August. But it's in the shadows at the edge of the lights that your chances of a hookup increase. That goes for tarpon, snook, seatrout, and whatever else is feeding.

I've learned to avoid fish in well-lit areas—although clearly visible, these fish are reluctant to strike. I work the edge of the shadows instead—the crisper the line, the better—and keep my fly in the dark. That's where predators are a lot less wary.

Stare into the water long enough, and you'll begin to differentiate between different levels of light. Where it's the brightest, light penetrates deepest; where it's totally dark, you can't see anything. Somewhere in between lies the kill zone, where making out shapes is difficult, unless you're a tarpon or a snook. Concentrate your efforts in this narrow band.

Docks attract predators with living forage—everything from shrimp to barnacles—with the emphasis on tiny baitfish. Docks with current are the most attractive. If snook are the major draw, then it's fly fishermen who benefit the most from these structures, especially after dark.

Sparsely dressed flies such as a size 8 white Estaz Minnow or Mike's Midnight Minnow fished on a floating line are deadly around docks after dark, although small, soft plastics and jigs also do a credible job. The gentle entry, coupled with a trim silhouette, does the trick.

Mike Conner fishes docks with his charters. As he puts it: "When the fish are targeting tiny minnows or baitfish no bigger than a dime, I use the smallest freshwater swim baits, like those made by Storm, I can find—those with just enough imbedded lead to allow for casting with light braided line. Otherwise, it's hard to beat a ¼-ounce D.O.A. shrimp, skipped under the dock."

I'll cover fish-fighting techniques in more detail later in this chapter, but as with fishing bridges, you're surrounded by pilings, so the goal is to keep any fish you hook from wrapping around them and cutting your line in the process. For that I'd

Snook in the dock lights. Fly fishing for snook under the myriad dock lights is a popular sport in both the Indian and St. Lucie Rivers. Seatrout and redfish share the glow with linesiders, as do bluefish on occasion. DAVID MCCLEAF

Mike Conner hefts a dock snook that hit in the lights. MIKE CONNER

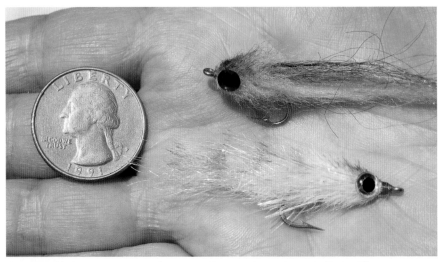

Tiny Midnight Minnow apes dock light forage (bottom). That's Conner's Glades Minnow on top.
MIKE CONNER

suggest either using heavy tackle or learning to steer fish into open water, which requires a mix of prayer and finesse. Remember the rule with snook? Hook one in the open and it'll head straight for the pilings, while one hooked next to the pillars bolts for the blue. Unfortunately, there's no guarantee, and there's probably more than one set of pilings.

Bridges

Man-made bridges form an integral part of the South Florida landscape. From tiny spans that connect residential side streets to concrete megaliths that vault the Intracoastal (ICW), all serve as magnets for aquatic life. This makes them important to shore-bound anglers. Bridges grant relief from the relentless current, while attracting forage and providing shelter. The position of pilings, and shadows at night, determine where different species gather.

Structures spanning fresh or brackish waterways attract a variety of cichlids, including butterfly peacocks, along with largemouth bass and one or more species of sunfish. Prime examples exist throughout suburbia. Peacocks, incidentally, prefer corrugated culverts that pass beneath those suburban spans. It must have to do with spawning terrain. You'll also find itinerant snook and tarpon, though tarpon move around more, rather than taking up residence beneath smaller spans.

Still, it's in salt water, not brackish, where bridge fishermen enjoy their finest hour—not only with nocturnal species that roam the shadows, but with schooling fish such as Spanish mackerel, bluefish, and pompano that approach certain bridges during daylight hours. Let's look at those daytime species first.

Schools of Spanish, pompano, and bluefish run through a number of bridges in Martin and St. Lucie Counties. Foremost among them are the Stuart and Jensen

Jimmy Duncan and family in days gone by, with a stringer of Spanish on the Lantana Bridge.
CAPTAIN JIMMY DUNCAN

A grouper joins this pair of trophy snook.
CAPTAIN JIMMY DUNCAN

Legendary snook fisherman Tom Greene with a former limit of four huge linesiders. TOM GREENE

Causeways between St. Lucie and Fort Pierce Inlets. There's a fishing pier under the Jensen Causeway where runs of mackerel aren't uncommon.

Two famous spans paved the way for this fishing. Little Blue Heron, at the eastern end of Blue Heron Boulevard in Riviera Beach, has traditionally been a mecca for Spanish macks. Spin fishermen by the dozens gathered on the lower of the two bridges and waited for the tide to drop. One of their favorite lures was a string of quills that they'd drag behind a PVC or bubble float or a piece of dowel. These splashy rigs were ungainly affairs, but they always brought up the fish. For the armada of boaters who anchored nearby, a single bay anchovy (glass minnow) on a long-shank hook was the bait of choice.

It started with a single note and turned into a symphony. The basin would remain quiet while the tide first ebbed. Then, an hour or two into the outgoing tide, the Spanish arrived en masse. They'd approach the bridge from the north—the upcurrent side on falling tide.

The free-for-all lasted for maybe an hour. Then, all at once, it would quit on cue. The cycle continued for several weeks, during March and April if my memory serves me. Something similar took place on Miami's Rickenbacker Causeway, when schools of Spanish swarmed into Biscayne Bay. The span connects Virginia Key (think Seaquarium and the Miami Marine Stadium) to the mainland. The macks, I've been told, have returned to the bay. Schools ran all the way from Jupiter to Fort Pierce Inlet in the Indian River a few years back.

Good daytime bridge fishing continues far up the coast, but that's beyond the scope of this guide. This you can take to the bank: Different species start sharing the limelight (along with the shadow line) once you get north of Lantana, with seatrout and redfish topping the list. You'll also find trophy black drum. Sheepshead become more common starting at Boynton, while flounder kick in from Jupiter north.

Then there's that irrepressible scamp, the toady (or blowfish), that should only be eaten if cleaned by an expert. (It's deadly poisonous if you break the wrong entrails.) Toady aficionados pursue them from September through December, when the blowfish run through Jupiter inlet. Go far enough north and you'll run into stripers (starting near Jacksonville). The well-known Stuart Crossroads—where the Indian and St. Lucie Rivers meet—is the farthest south a striped bass has been reported.

Night Fishing

Anglers pursue both snook and tarpon on South Florida bridges at night, along with—at one time—Goliath grouper. The farther south you go, the more tarpon predominate—as is usually the case when you head farther inland. However, all three species are available in inland waters from Sebastian Inlet to Biscayne Bay. Still, snook fishing remains the nighttime tradition.

At one time, South Florida held limitless possibilities for fishing on foot from a bridge. Now, some new spans soar so high that fishing is uncomfortable or, worse, impossible. In the words of George Copeland: "Some are so high that your mullet gets a nosebleed. And the railings are too low to lean against."

A number of bridges have been declared off limits because of increased boat traffic and safety concerns. Liability trumps recreation when it comes to fishing—a theme we keep hearing from public officials. The risk of injury is real, however, since cars wreck on bridges and motorists throw objects from cars. It's safest to fish bridges where the catwalks are separated from vehicular traffic by height, distance, or screening. You can also fish an old bridge that's been closed but portions were left standing—like the Sunshine Skyway over Tampa Bay or several spans in Martin County.

I've fished from various bridges since I was a boy. I caught my share of snook after mastering the basics, but certain fishermen always outdid me. Tom Greene, George Copeland, David Justice, and others now enjoy immortal status fishing for snook from bridges. What they learned in the shadows has helped us all. Greene helped me with this chapter, providing a few tips on how to get started bridge fishing.

Tides

To Greene and Copeland, who fished together, tides were a religion. They'd grab the first thirty minutes of the rise at Flagler (in downtown West Palm Beach) and then hightail it south on U.S. 1 to Lantana Road, where they'd fish for a few more minutes—and so on.

Tides along Florida's southeast coast vary only slightly from inlet to inlet because the inlets are arranged along a nearly straight line that runs north and south. However, inside those inlets—as well as in the Intracoastal—the timing differential increases just as the tide works its way up a salt marsh creek.

Small differences in location translate to hours with tides, so you need a reliable tide table: one that lists not only highs and lows, but also the height of each, along with lunar data. Go online (or check the resources in the back of the book). Tide data is critical when you're fishing bridges.

Current also factors into bridge fishing. Its direction depends on your position between the two closest inlets—the outgoing tide flows toward whichever inlet is closer.

But there's a bit more to it. Say you live in Fort Lauderdale, with its miles of canals. The nodal point between Port Everglades and Hillsboro Inlet isn't halfway between the two—not with all that water in Lauderdale proper. So the midway point for tides is closer to Commercial Boulevard—nearly two-thirds the way from the port to the lighthouse.

The position of the shadows determines whether a bridge is an incoming or an outgoing span. Snook typically bite better on the outgoing tide, but nothing in fishing is etched in stone. Greene and Copeland kept moving. Dead-slack high, when the current comes to a halt, was hard to beat for Goliath grouper—back when it was legal to possess one.

Something else about tides: They're stronger during and immediately following full and new moon periods. Plus, an onshore wind helps them rise higher than usual—and impedes their fall. This can either add or subtract from current and water depth, which has a direct effect on where the snook are lying.

Bridge Tackle

You'll find plenty of places to indulge your love of light tackle. Under a bridge isn't one I'd suggest, although you can cast plugs or bucktails if you're familiar with the span and your tackle.

Snook fishermen say that if you hook one in the open, it heads for the pilings, while one hooked between them does just the opposite. But sooner or later, they all look for obstructions. As I mentioned, although snook aren't known for endurance, when they turn on the gas it can really scare you. While a longer rod helps position your bait, it can work against you through something engineers refer to as mechanical advantage.

Stay on the alert for feeding activity: crashes, pops, or skittering baitfish. If the pops appear close to the pilings, you'll need to haul out your heavy tackle. In bridge parlance that means a 10-foot rod fitted with a solid spool reel that's filled to the bars with 60- or 80-pound mono. The longer rod helps position your bait, while the reel, essentially, works as a clutch. As far as the drag, you want minimal slippage. A 25-pounder on a 10-foot bridge rod and 80-pound line will pull a man off his feet—and drag him along a concrete railing.

A celebrated high school wrestler and power lifter I knew used to crawl underneath the Ermine River Bridge on U.S. 1 in North Palm Beach. His goal was to yank monster snook from a space so small that he could barely stand up. He achieved it. The Ermine was known for its trophy snook, and the fish held close to the pilings. So my pal tossed his mullet directly between them. Then, with a mighty heave—he used 100-pound-test—he'd launch his victims straight up on the rip-rap.

His pool cue of a rod measured 4 feet long—it was built like a kite rod with one or two guides. Yet it worked like a charm in confined quarters, where snook fishing was more of a contact sport.

The last I heard of him was from a mutual friend who'd seen him fishing the Lake Worth Spillway, which has lots of big snook when the floodgates open (see page 142). He was dangling a giant shad from another derrick, using a reel that was filled with Weed-Eater cord (or a reasonable facsimile). The line was too coiled for the shad to straighten it. But when asked about his no-nonsense setup, he said, "I skeeze 'em." He was grinning from ear to ear.

Live Baiting

The basic method for dangling live bait is free-lining a mullet, whitebait, mossbunker, shrimp, croaker, or sand perch under the section of span where you think the snook are lying and then moving if the fish quit hitting. Look for sharply defined shadow lines.

Baits can be hooked through the nose, the upper lip, or slightly ahead of the spiny dorsal. Shrimp can be hooked in the head (just ahead of the black spot) or tail. A trick I learned from Tom Greene is to hook a mullet or sand perch just inside the vent—lightly, so as not to injure the bait. This encourages it to swim down toward the bottom.

Tom Greene's Bridge-Fishing Tips

Scout the scene. The first thing you do when you get to a bridge is look around. Look for shadow lines and the presence of bait: mullet, whitebait, or shrimp. "April through July is the best on bridges," Greene advised me, "along with October, during the mullet run." If possible, talk with the bridge tender: He'll say right off the bat if you can fish the draw span, where you'll usually find a no-fishing sign.

The shadow knows. "More than anything else, think shadow lines, which—when you return to fish—should be on the upcurrent side. That depends on where the streetlights are located, so one bridge may fish better on the incoming tide, while the one next door fishes just the opposite." The middle of a span isn't always the best place to fish. Look for the crispest shadows and moving water, which could be next to the seawall.

Work as a team. "Try not to go it alone," Greene says. "We traveled in threes whenever we could. We'd let one guy off at the foot of the bridge; then have him walk it before picking him up. That way we'd get an accurate report."

"When we were kids," Green added, "here's how we'd fish bridges [meaning long ones over the Intracoastal]: Once we picked our spot, we'd have the driver drop us off with our tackle, before finding a place to park the car. Then, when we got ready to leave, we'd reverse the process, after making sure that the streets were empty." That gave them time to load up the car on the bridge.

Monitor the tides. "You won't believe how many people don't realize that a mile or two in the Intracoastal can mean an hour's difference in the high or low." Greene says. "Or that the snook under a particular span may only hit during a very brief stage of it."

Use good sense and be safe. "Look for a parking space at the foot of the bridge, so it's easier to keep everything in sight. Fishing at night always involves risk, so rely on your two greatest allies: darkness and a cooperative bridge tender."

Obtaining live bait is critical for bridge fishing. Broward has a 7-foot limit from collar to lead line, a move intended to curb commercial fishing. Of course, you can always buy bait (especially shrimp) or catch whitebait with a Sabiki on a local pier. PAT FORD

When snook are feeding on top, you can pick out their shapes on the leading edge of a shadow line, but sometimes they lay farther back or, as Greene reminded me, near the bottom.

Remember the current? When your line comes tight at the end of a drift, reposition your bait with a flip of your rod to a spot just ahead of the shadow line.

South Florida Bridges

There are hundreds of bridges in South Florida, many of them spanning the Intracoastal. Here are a few of the better ones. When I refer to these bridges by name, I'm referring to actual streets or highways. The term "causeway," on the other hand, means "elevated road."

In the following travelogue, Tom Greene and friends take us from Biscayne Bay north to Fort Pierce Inlet. Most bridges in between have been omitted, due to a lack of space.

Miami-Dade County

Bridges that connect Miami Beach with the mainland are popular with the local fly-fishing guides, who use skiffs to position their clients near the tarpon and snook that work the shadows. Hot spots include the 36th and 79th Street Causeways and several smaller spans.

79th Street Causeway. At one time, patrons at Mike Gordon's Seafood Restaurant, located on the northwest corner of the bay at NE 79th Street, tossed oyster crackers to schools of hungry ladyfish. Then as now, the best fishing in that part of the bay takes place during the first few hours of the outgoing tide. The best spot is from a skiff. That said, the late Harry Friedman and Joe Brooks, who stayed with Friedman at his home in Miami Beach, fished with fly rods from bridges after returning from evenings on the town.

Broward County

Fort Lauderdale—the self-billed "Venice of America"—is laced with miles of inland waterways. Bridges are legion, but not all are good fishing spots. In days gone by, several spans earned reputations for tarpon and snook, such as the 17th Street Causeway and the five bridges over the Mercedes River. Both were snook hot spots but are now strictly off limits, as is the 7th Avenue Bridge over the New River, which was great for tarpon. The new bridge over 17th street is too high to fish from. However, there are still some bridges open to public fishing.

Las Olas Boulevard. This one's still fishable, and according to Tom Greene, it's a good bet on the outgoing tide. He referred to a shadow line on the northwest corner—just east of the Las Olas Isles. I lived in the islands in the early 1970s but I never fished there until later. When I did, I enjoyed some action with tarpon, snook, and jacks.

The Two Sunrise Bridges. You'll find two spans near the beach on Sunrise Boulevard: one that crosses the Intracoastal and a lower version over Middle River.

Fort Lauderdale's Little Sunrise crosses Middle River. STEVE KANTNER

I've landed countless snook at Big Sunrise. However, the basin to the north is now so crowded with mega-yachts that I question if there's still any current.

Little Sunrise over Middle River always left me flat. However, the bridge a ways upstream, dubbed "Jefferson Bridge" after the super-store that abutted it on North Federal Highway, was a different story. Greene, who referred to the span as reliable, was crawling beneath it while retrieving a snook, when he slipped and fell on the seawall oysters and received severe lacerations. Wounds of that nature are extremely dangerous and require immediate medical attention.

Commercial Boulevard. It's still possible to fish here, according to Greene. During the early 1970s, I hooked a snook in the channel that would probably have been my all-time record, but there was no one there to fetch a bridge gaff. I attempted to free-spool it around the fender. But when the fish refused to cooperate, I attempted to lift it—a joke on 60-pound line.

Atlantic Boulevard in Pompano Beach. Every bridge has idiosyncrasies, this one included. Tom Greene suggests parking at St. Martin's Episcopal Church on the southwestern corner of Atlantic Boulevard and dropping a live mullet back under the bridge. The outgoing tide runs north at this span, and tarpon feed along a well-defined shadow line.

18th Avenue in Pompano Beach. This one can be good, according to Greene, if the Cypress Creek locks (C-14) are open. I've heard stories of giant bull sharks swimming this far inland while chasing mullet.

14th Street in Pompano Beach. Greene remembers how sand perch would follow the seawall north on the outgoing tide. After netting a bait, he'd tail-hook and free-spool it and allow it to drift 100 yards or more, while steadfastly anticipating that telltale thump. This is one time you shouldn't fish the shadow line, but should allow your bait to swim with the current. The lesson here is to experiment.

Hillsboro Boulevard, or Butler Bridge. One of Tom Greene's favorites—apparently, best on the falling tide—and applauded by others. This bridge is located a few miles south of Boca Raton Inlet. The last time I crossed it, it was still open to fishermen.

Palm Beach and Martin Counties

The snook here are larger and more accessible, with multiple venues including the surf, spillways, inlets, and piers, each contributing to a strong population. Check out the long, low bridges that span Lake Worth Lagoon, where there's less pollution from feeder canals.

Camino Real Bridge in Boca Raton. Renowned architect Addison Mizner envisioned a Royal Road connecting his Cloister Inn (now the Boca Hotel and Club) with Henry Flagler's railway station.

When I was a kid, a teenager named Bobby Schactel landed a *Field & Stream* magazine merit-badge snook while standing on Camino Real Bridge—in the middle of the day. He saw jacks harassing some mullet and cast his plug down the seawall. If my memory (or Tom Greene's) is accurate, it was a Heddon Baby Zara, and the snook weighed 41 pounds. Schactel was one of our boyhood heroes. (He was also the official shark fisherman of Deerfield Pier, whom the management called to remove nuisance predators.)

I fished the Camino Real Bridge on summer nights, and it was one of my favorites. Tom Greene, who grew up down the street, knows it better than anyone.

The year Tom graduated from Boca High, the senior prom was held at the Boca Club. During a break in the festivities, Greene and his date took a stroll down Camino Real and onto the bridge.

Tom saw a dozen fishermen baiting snook in the shadows. Forget that he was wearing a tuxedo or that he was with a date—he wanted in on the action. After borrowing a rod from an angler he knew, he tossed a live mullet beneath the span. When he felt a thump, he leaned back hard and flipped the 15-pounder onto the bridge, before dutifully relinquishing the rod to its owner, along with the fish.

Greene admitted once that, in a moment of weakness, he missed his sister's wedding—where he was scheduled to be an usher. The fishing was so spectacular at the spillway, he claimed, that he forgot to look at his watch. You can read the full account in his book *A Net Full of Tails.* Like that kind of fishing, it's too good to miss.

Boca Inlet (A-1-A). Tom said he's fished this bridge since back in the days of the old wooden structure. I have, too. It attracted hordes of snook that ate jigs on the falling tide, along with monster jacks and tropical reef species. A bascule bridge now spans Boca Inlet, a natural watershed that was originally dredged in the 1920s so that yachtsmen could access the Cloister Hotel.

Palmetto Park Road. From his earliest days at Boca Tackle, located down the street, this was Tom Greene's home water.

Tom said one of his heroes lined the span in the evenings waiting for the tide to change. They cast jigs back then—long, white Nylures—but not until the tide started moving. One reason they caught snook without losing lures was the sandy bottom that lies north of the bridge. Jigs remained popular throughout the 1960s, when live-bait techniques became more refined.

Schools of bait rushed through the span at the start of the outgoing tide—a mixture of mullet and Spanish sardines. But the regulars ignored the ensuing commotion, as snook exploded through the panic-stricken schools. They kept casting their jigs. Except for Tom.

This was the start of his live-baiting career and when he learned to throw a cast net. But more than snook lurked in the shadows.

I remember a tattered snapshot of a grouper that a guy yanked from that span on a rope. The monster weighed 350 pounds. According to Tom, a wrecker had to hoist it over the railing. A 7-foot moray eel was yanked from the span, too. According to scuttlebutt, it weighed over 100 pounds.

Atlantic Avenue in Delray Beach. A "keeper" according to Tom. This bridge is wide with plenty of current. It also has a good shadow line.

Lantana. This bridge is the first to cross Lake Worth north of Boynton Inlet. It's located next to the charter docks. Greene remains partial to Lantana because it sits near Boynton Inlet and the Boynton spillway, where snook always gather when the floodgates are open. The bridge is close to the inlet, assuring that there's plenty of current.

Lake Worth. Greene says you could (and probably still can) catch snook there during daylight hours, especially on spans where the shadows are wider. With fewer docks and canals for shelter, the snook seek relief from the blazing sun there during July and August when temperatures soar.

Tom would leave Lake Worth Pier with a live well filled with whitebait, dork jacks (baby blue runners), or any available baitfish and free-line them in the shadows beneath the spans where, with nowhere else to hide, the snook were waiting.

Southern Boulevard. There's an old story about a bridge fisherman landing a tuna here: a yellowfin, I recall, of considerable size. The tuna was supposedly alive, yet the fisherman snagged it. I'd think it had to have been sick or injured.

You'll see a smaller span as you approach Palm Beach Island, and I saw people fishing it recently. To get to either span, take the Southern Boulevard exit east from I-95.

Royal Palm. The middle of the three Palm Beach bridges. To get there, take Okeechobee Boulevard east from I-95. The bottom here is covered with grasses, improving your chances of finding fish—jacks, ladyfish, pompano, bluefish, and even an occasional permit.

Flagler. The northernmost Palm Beach bridge is Flagler. George Copeland and Tom Greene would only fish it for thirty minutes before heading south to follow the tide. In addition to the nighttime snook activity, I've caught pompano, ladyfish, and bluefish in the Lake Worth Lagoon between Flagler and Southern Boulevard, not all from a skiff. Quite a few of those bluefish were trophy-size, upward of 15 pounds.

Blue Heron. The first bridge north of Palm Beach Inlet is Blue Heron. The big span's too high; as for the other, we'll have to wait and see. (It's currently under repairs.) Before, it yielded mackerel, blue runners, and barracuda.

Indiantown Road. I used to pass this on my way to work, but I never took time to study the shadows. The Burt Reynolds Park boat ramp, a short distance to the

Flagler Bridge, over Lake Worth Lagoon, earned a reputation for large snook. STEVE KANTNER

north, teems with baitfish. I've seen tarpon and jacks while launching there, so the bridge, I'd bet, captures some spinoff.

Jupiter Inlet. Portions of the old A-1-A Alternate Bridge have been left standing to serve as fishing piers, and people fish there during the day. At one time, you could catch snook, as well as jacks and some monster bluefish, from this bridge.

708/Bridge Road. Judging from the mangroves, this bridge should be a winner. Mike Holliday, former editor of *Florida Fishing Weekly* tells a story about this span that I find amusing: He'd been dropping live baits and fishing the shadows. (It's quite a ways to the water. The best he could do were several missed strikes, and Mike could write volumes about setting a hook. All he got back, though, were simply empty; plus, his leader was never frayed. Frustrated, he decided on a different approach: pulling gently when he felt the crunch. So what eventually appeared in the bridge lights? A fat, brown otter holding onto Mike's mullet with its powerful claws. After Mike retrieved it—the mullet, that is—he tossed it back to the hungry varmint before gathering his gear and heading elsewhere.

Mike shared another anecdote:

I caught my largest snook ever from the Hobe Sound Bridge on a live 12-inch mullet. That was on the old bridge, and I was fishing at night with straight 100-pound monofilament and a rod you could make a pool stick out of. I hooked a fish that ran out from the bridge and then made a big bow and headed back for the pilings. The 6:1 gear ratio on a Shimano Speedmaster IV, which at the time was the newest, most progressive reel on the market, let me catch up to the fish and steer it away from the pilings.

While fighting the fish, a truck stopped on the bridge behind me. Once I had the fish subdued, I could look back, and it was a Florida Marine Patrol Officer in the vehicle. He asked me, 'What do you have on the line?' to which I replied, 'The biggest snook I've ever seen!'

He pulled his truck to the end of bridge and parked, came up on the bridge and helped me land the fish using a bridge gaff, and then we weighed the fish on his hand scale. It weighed 42 pounds.

Those big fish taste like crap, so I let it go.

Stuart

Here, several older spans have been replaced by new ones that vault both the Indian and St. Lucie Rivers.

Stuart Causeway. The 10-cent and 25-cent bridges—names dating back to when the bridges had tolls—are now a part of the Stuart Causeway that crosses both the St. Lucie and Indian Rivers. Shore-bound anglers have limited access to Spanish mackerel, bluefish, and pompano at specific times during the winter when the water's not too cold and there's plenty of forage. The easternmost span is a favorite for pompano. The locals refer to it as a mosquito bridge.

Snook fishing is also good at times; plus, I once saw a nearly 14-pound seatrout landed at the old 25-center, and that was close to the all-tackle world record. If my memory serves me, it hit a live finger mullet at daybreak. Captain Steve Anderson offered the following: "As for big fish, it's snook, tarpon, trout, and Goliath grouper;

Jensen Beach Causeway combines height with utility; no need to open a draw span. MIKE CONNER

This fat, river pompano hit a Goofy Jig. PAT FORD

as for smaller fare: black drum, sheepshead, snook, pompano, croakers, sea trout, bluefish, ladyfish, and sand perch."

Keep in mind that the 10-cent and 25-cent bridges have both been replaced by a newer span, with exits at Sewell's Point, where the St. Lucie and Indian Rivers meet. To get to the causeway, take the Stuart exit (State Road 706 or Kanner Highway) east from I-95; then follow it all the way to Indian Street. Take a right on Indian and continue to the end. You'll cross both Federal and Dixie Highways. Then bear to the left on St. Lucie Boulevard. St. Lucie winds through a residential neighborhood before reaching the foot of the causeway.

You'll find nothing at this causeway like the catwalks of old—except at the western end of the former 10-center, where people fish from the bulkhead. I've been told, however, that this is private property.

If you're headed east, there's another option near the eastern shore of the Indian River. Circle back around and head toward the boat ramp (at the southeast corner of the causeway), and you'll cross a short, low bridge. That's one of the mosquito bridges where anglers stand on buckets to jig for pompano. It's a proven producer on the outgoing tide.

Jensen Causeway. A few miles north of the Stuart Causeway is Jensen Causeway, which crosses the Indian River. You get there by taking A-1-A north or by following U.S. 1. You'll find the same species here as you do farther south. Worth mentioning is that anglers on the easternmost span (another of the mosquito bridges) experience reliable runs of pompano.

Steve Anderson shares his fishing methods for the Jensen Causeway, either winter or summer. He suggests using

> feathers [a red-tailed hawk] retrieved against the last of the incoming tide—for snook. Or use a live throat-hooked silver mullet, and toss it up-current while maintaining tension [tuck] on your bait, which forces it to go deep. When the mullet approaches the shadow line, ease off on the tension and let it swim naturally.

A red-tailed hawk is a white jig with a long, red trailer. STEVE KANTNER

Nighttime is the right time when fishing docks and bridges. PAT FORD

You can also cast feathers along the shadow line, when the tide slows in either direction. A live bunker [menhaden] hooked behind the head or throat-hooked can be fished the same as a silver mullet.

A jumbo live shrimp fished a few feet off the bottom behind a trolling lead (3 to 6 ounces) and 4-foot mono leader can be walked slowly up and down the shadow line as soon as the tide slows down.

Roosevelt. This bridge is on U.S. 1 between Stuart and Fort Pierce. A favorite with snook fishermen, it crosses the St. Lucie River where it turns to the west. Due to sporadic freshwater releases, the fishing runs hot and cold.

Anderson says of the Roosevelt: "During the daytime, I've caught snook, trout, drum, sheepshead, flounder, jacks, bluefish, mangrove snapper. At night, it's snook, trout, Goliath grouper, ladyfish, bluefish, drum, and mangrove snapper."

Mike Connor added that Roosevelt is actually two bridges: "There's an older Roosevelt, which is the lower of the two, where fishing is permitted over a guardrail. Plus, there's the newer Roosevelt, which is a higher bridge—with an excellent catwalk at water level—that originates at Flagler Park. The park is located in downtown Fort Pierce on the southwestern shore of the St. Lucie River."

Anderson shared a tip about Roosevelt: "I fish the outgoing tide the same way—across the highway on the west side of the bridge. You can fish most of these bridges on the slack for jewfish (okay, Goliath grouper) with a live or dead big bait fished on the bottom behind the catwalk. Rig a sliding sinker 3 feet from the hook, and be prepared for big fish!"

Anderson also mentions that the U.S. 1 Cut-Off Bridge has fishing for snook, drum, sheepshead, jacks, and bluefish during daylight hours. "The South Bridge during the daytime," he said, has "snook, drum, sheepshead, flounder, snapper, pompano, mackerel, bluefish, ladyfish, jacks." At nighttime, you can find snook, drum, croakers, sand perch, snapper, flounder, grouper, ladyfish, and seatrout.

"The North Bridge at Fort Pierce during daylight hours," he said, has "snook, drum, sheepshead, flounder, grouper, snapper, trout, jacks, croakers, pompano, bluefish, ladyfish; at night, the same."

Inlets and Jetties

Ocean inlets—or better, the jetties that frame them—provide low-cost fishing for a multitude of species ranging from snook and tarpon to run fish, along with oversize snappers and a few surprises. Admission is free or inexpensive, making them viable alternatives to owning a boat.

Jetties redirect the longshore current, thereby preventing a buildup of sand. This keeps inlets open and helps combat coastal erosion.

The new south jetty at Jupiter Inlet. The list of species available on jetties depends on tides and time, with large predators preferring the cover of darkness—at night or in off-color water.
STEVE KANTNER

Jetty Basics

The best jetties to fish from are the ones in the path of migrating fish: mullet, mackerel, bluefish, snook, pompano, or anything else that moves with the current. That's the north jetty in the fall or winter; the south in the spring. It may seem unusual, but the one just across the inlet may be totally devoid of migrating fish.

The fish we catch from jetties are there for a reason—typically based on the tide. While the cast of fish keeps changing according to a schedule, the players take their cues from the tide's rise and fall. So when you hear of the capture of a deep-water species, such as bonito or kingfish, it probably took place at the top of the tide. That's also when whitebait are present in numbers.

High tide is when inlets run their clearest, which is what offshore species prefer. I've heard plenty of tales about Palm Beach County, where blue offshore water comes the closest to shore. In fact, just last summer Terry Luneke landed a bonito on the Jupiter jetty in crystal-clear water. The north jetty at Sebastian Inlet—although it's beyond the range of this book—used to be famous for runs of kingfish.

Low tide attracts inshore predators such as snook and tarpon, which are typically more active during low-light periods (after dark, or in off-color water). As

Every jetty has recognized hot spots that typically change with the tide. Rocks, ledges, or other features on the bottom deflect the force of the tide. STEVE KANTNER

soon as the tide in an inlet starts falling, the water begins losing its clarity. How opaque it gets depends on what's behind it—be it effluent from the Intracoastal or water from the state's interior. Both are byproducts of any recent rainfall.

Some species pass through inlets en masse, a phenomenon influenced by the time or the phase of the tide. Take the schools of snook that appear out of nowhere as soon as the tide starts to drop. The exact timing varies from inlet to inlet.

When it comes to water color, jetty fishermen recognize two distinct variations: green and brown. Green means salt, while brown signifies fresh. During incoming tides the surf gets roiled (think pea soup). Keep in mind, however, that opacity (cloudiness) and color (a measure of freshness) are different and that every species has its preferences.

Both opacity and color are open to interpretation; plus, outgoing tide provides cover like nightfall. Then, falling water tends to concentrate forage.

Forage must be present to begin with, and it enters inlets on the incoming tide. When the tide starts falling, the forage exits—while predators await it in the off-color water.

Inlet Species

Once the rainy season begins around the end of May, freshwater runoff changes the salinity in inlets and provides ideal conditions for snook to spawn—and that's what draws them.

Down south where I live, during May and June, fishermen hook the majority of snook on the channel side during nighttime outgoing tides by winding crosscurrent at a steady pace. Quarter your lure upcurrent before allowing it to sink. You'll do better on the beach side when July rolls around, especially on an incoming tide, when snook stage there prior to spawning.

One of the easiest ways to catch snook at night in inlets is with a weighted grub on spinning or plug gear with 10- to 20-pound line. During the daylight hours you'll do better with bait.

Summer's the perfect time to enjoy catch-and-release fishing. Fly fishermen can share this resource by casting full-bodied streamers, size 2/0 to 4/0, on sinking-tip or sinking lines. A white Deceiver is one of my favorite patterns. Use a monofilament leader of 40- to 60-pound-test with all types of tackle.

Favored live-bait offerings in inlets include live thread herrings, Spanish sardines, pinfish (sailor's choice), and croakers—all fished near the bottom with a lightweight sinker. Sailor's choice and croakers work better in August. There are times, however (in winter, for example), when you can't beat a live shrimp fished on a troll-rite—just a jighead without a skirt. Fish all these offerings near the bottom or slowly retrieve.

Tarpon are another inlet favorite, though the odds of landing one are less than they are from the beach. The best fishing takes place in the fall, when both species of mullet enter the inlets, and then again in spring, when the silvers run. Tarpon also gather in inlets during midwinter shrimp runs.

When the mullet run, they gather alongside jetties and wait for the outgoing tide. But it's at dead-slack high tide that the tarpon start driving panic-stricken mullet against the rocks. This is the time for a nose-hooked, free-lined live mullet. At Palm Beach, Fort Pierce, and Sebastian Inlets, permit join tarpon along the current seams and pick forage from the drift lines, mostly at night and when tides are the strongest. Crabs, which come up and swim on the surface, figure prominently in the forage du jour.

Pompano, which typically run in the surf, enter inlets (Palm Beach, Jupiter, St. Lucie, and Fort Pierce) during the winter. Once inside, they head for grass and shell flats or channels, where you'll occasionally find them shadowing manatees. Sailfish Flats, near Jensen Beach, is an excellent example of this type of habitat, as are shoals farther north in the Indian River. The schools, which spook easily, move with the tide.

One of the easiest ways to locate pompano is to look for them skipping. They skip out of the water and land on their sides, usually in response to boat wakes—a visceral response to being disturbed. My love for this species is rooted in the kitchen, and I'm not alone in that.

You can find permit in inlets as far north as Sebastian. Drift a small crab, preferably at the start of a falling spring tide.

While it may not happen as often as it used to, kingfish and sailfish still enter inlets—if there's a draw (say, the incoming tide or an abundance of baitfish). I narrowly missed seeing a sail reeled to the rocks on—of all places—the John U. Lloyd State Park jetty.

Redfish are a draw on several jetties, and their numbers increase as you go farther north. While inshore populations exist as far south as Munyon Island (in North Lake Worth), it's not unheard of for reds to migrate south. They typically do this in November in the wake of a strong Nor'easter. It's anybody's guess where they eventually end up, but I'd bet in either Jupiter or Palm Beach Inlet.

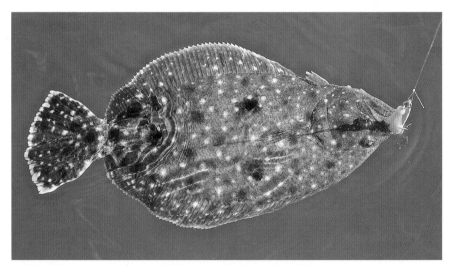

Southern flounders draw crowds during winter. The best way to hook those 4- to 10-pound doormats is to drag a nose-hooked live finger mullet or other small baitfish slowly across the bottom up and down the jetty—preferably rigged behind a sliding egg sinker rig. PAT FORD

Bluefish love the surf and off-color water. They push through inlets on the incoming tide—especially from Palm Beach north—if conditions in the inlet resemble those along the beach or if the surf gets too cloudy. PAT FORD

Large bull reds strike baits fished on or near the bottom (a live finger mullet rigged with a piece of clam is a favorite in Mosquito Lagoon), but they also hit lures intended for snook. Large jigs are a favorite, as is a live shrimp fished on a troll-rite, but there's still no substitute for a live finger mullet or a crab.

You can still catch reds from the Fort Pierce jetty. I landed one there on my very first cast and landed another from a skiff near the Sebastian jetty that weighed 35 pounds. I hooked a second fish that stripped the reel. Reds surge through Sebastian on the outgoing tide.

The mainstays on jetties are ordinary panfish, including Spanish mackerel, sand perch, pompano, various snappers, occasional small groupers, sheepshead, and a host of others. Also popular are the jacks and blue runners that strike small jigs and spoons regardless of the tide.

Most smaller jacks are seasonal migrants whose numbers increase in fall. Because glass minnows collect around jetties, carry a light spinning outfit rigged with a quarter-ounce white marabou crappie jig.

Snappers—especially mangroves—can be found around jetties, with the best fishing taking place during summer and fall. Of particular interest are the large snappers that enter inlets before spawning. Look for spawning mangroves at Jupiter Inlet near the north jetty starting around the first of September. Meanwhile, it's not unusual to catch cubera snappers while fishing live bait on the bottom or dragging a deep-running plug, such as the Rapala X-Rap. Like the big spawning mangroves, they average nearly 10 pounds apiece.

Groupers, like snappers, seek shelter in the rocks. In fact, there was a thriving fishery inside Palm Beach Inlet, where gags accounted for the majority of catches. But that fishery, as we knew it, is gone now. Goliath grouper are currently under federal protection, so hopefully the fishery will have a chance to renew.

Spanish mackerel are another visitor—not only during the fall and winter, but whenever schools of glass minnows show up near jetties. The methods for catching them are the same as at piers. Bluefish are another jetty favorite.

South Florida Jetties

South Florida has nine major inlets (Fort Pierce, Stuart, Jupiter, Palm Beach, Boynton, Boca, Hillsboro, Port Everglades, Haulover)—not counting the channels between the Atlantic and Biscayne Bay, only seven of which have jetties that are even remotely accessible, usually one per inlet.

Jetties vary widely from dilapidated rock piles to modern, paved breakwaters with railings. The thing about jetties, no matter where you find them, is that the fishing changes as fast as the tide. Please keep that in mind while you use the following information.

Haulover Park and Marina

Baker's Haulover is a man-made inlet in northeastern Miami-Dade County. Tom Greene, who fished there years ago, claims that he and his friends caught plenty of snook there. The current buzz is about the nude beach in Haulover Park.

The John Lloyd Jetty

The south jetty at Port Everglades Inlet, at the northernmost end of John U. Lloyd State Park in Broward County, has always been a hot spot for snook and pompano. It just reopened to fishing after extensive repairs. To fish it at night, you need a jetty stamp, which the park staff will add to your annual permit bringing the total price to $60. The permit is also good in other state parks—a bona fide value. The north jetty—however forbidding—probably continues to concentrate migrating gamefish.

Back when I fished the John Lloyd jetty, we'd scramble to reach a particular boulder before casting into the channel on an outgoing tide. The snook were so thick that we'd bump their backs. But every cast or two, they'd grab our grubs, which resulted in multiple hookups. Then the fish would quit, as if someone threw a switch. Since they'd always stop at exactly the same time, we assumed they were moving en masse—either at a particular stage of the tide or when the light diminished.

We never got the chance to prove our theory: Park rangers held us at the gate—until those magic minutes had nearly passed—while they cleared the park of any daytime patrons.

I hooked snook on fly gear by dredging with Scientific Angler Wet-Head or Wet-Belly lines and large, white Deceivers that I cast on the beach side, where the current was slower.

Hillsboro Inlet

When I was a teenager, I had a key that gave us access to the tattered south jetty via Wahoo Bay. Now the bay is gone and so is the key, meaning the only way to get to Hillsboro Inlet is by hiking the beach from a tiny parking lot at NE 14th Street.

The north jetty is unfishable because of gaps in the boulders, and aside from that, it sits on government property. My fondest memories of the old south jetty were of ballooning out for sharks (see page 180) and drinking beers with my girl-friend, not necessarily in that order.

Boca Inlet

The south jetty is part of South Inlet Park. To get there, take A-1-A north from Hillsboro Boulevard (which exits on I-95). The north jetty is no longer accessible due to property restrictions. Fishing for snook and Spanish mackerel is still reliable. Not that long ago, when a friend and I walked from the bridge to the old north jetty, we caught our limit of snook on 77M MirrOlures on the falling tide.

South jetty at Boca Raton Inlet. STEVE KANTNER

Boynton Inlet

Boynton Inlet was famous for snook and big jacks at one time, and both jetties have been repaired and are open to fishing. The stretch of coastline immediately surrounding it comes closer to the Gulf Stream than anywhere else in North America.

To get to Boynton Inlet, take A-1-A north from Delray Beach or I-95 south from Palm Beach. Be careful driving to the north parking lot. The mansions in this area are surrounded by vegetation that sometimes encroaches on the road.

You can get up-to-the minute information on the tides and weather, as well as a webcam, both in Palm Beach and Martin Counties from Erdman Video Systems. (See the resources in the appendix.)

Palm Beach Inlet

The north jetty at Palm Beach Inlet is a great choice whenever the surf is too rough or the water at the piers is choked with sand. To get to the north jetty, take a right at the first light east of the Little Blue Heron Bridge (on Blue Heron Boulevard). Then follow the signs to the Palm Beach Shores City Hall, which usually has ample parking. Walk south to the inlet seawall and follow it straight to the jetty.

Boynton Inlet from the north parking lot. STEVE KANTNER

Boca Inlet, looking northwest toward the A-1-A bridge. STEVE KANTNER

Deep inlets like Palm Beach that accommodate seagoing vessels have powerful currents. In days gone by, anglers on the south jetty would wait for the mullet to make an appearance. Snook and jacks tore through the schools with fervor, while the Palm Beach Jetty Conchs made some incredible catches. Now, limited access has shifted the focus to the short north jetty and north inlet seawall. Still, if you're adamant on fishing that famous south jetty—where parking is currently nonexistent—take Bus 41 from West Palm Beach.

Tom Greene fished there in the past with bottle caps screwed to the soles of his shoes. The fishing, he told me, was unbelievable. Of more recent interest are the jacks and bluefish that gather across the inlet at the start of the outgoing tide. Sharks, supposedly, are also reliable. Count on Palm Beach Inlet under certain conditions (meaning when conditions are wrong everywhere else).

Fort Pierce Inlet

Fort Pierce is another deep inlet that hosts quite a few tarpon, along with flounder and redfish and quite a few snook. The pompano runs are also reliable. Anglers fish for tarpon with huge wobbling plugs (Rapala X-Rap) that they toss from the jetty tip and dangle in the current. You don't need to reel; just cast and hang on. Once again, do this on the outgoing tide. Anglers can access the jetty via Fort Pierce Inlet State Park. To get there, take the beach road north along Hutchinson Island.

APPENDIX: RESOURCES

Florida Highway Patrol: Dial *FHP

Florida Fish and Wildlife Conservation Commission
For laws, limits, and licensing information: www.myfwc.com
To report violations:
    Phone: 1-888-404-3922
    Cell: *FWC or #FWC
    Text: Tip@MyFWC.com

Maps of Water Management Canals: www.state.fl.us/fwc/fishing/pdf

Current Regulation Changes: saltwater-l@listserv.myfwc.com

Tidal Data: www.saltwatertides.com

Rods, reels, and tackle: Can we ever have enough? While certain basics are essential, it's nice to indulge a whim or two. PAT FORD

Real-Time Weather: radar.weather.gov/radar.php?rid=mlb

Fresh and Brackish Water Temperatures and Flow Rates:
waterwatch.usgs.gov/new/?m=real&r=fl
(This site is particularly useful for Florida Bay and salt marsh rivers.)

Spillway Information:
www.sfwmd.gov/portal/page/portal/levelthree/live%20data
(Click "Water Control Gates" on the left of the page, before scrolling down
and entering the name of the structure. Under "Parameters," click "discharge.")

Groundswell Predictor: www.magicseaweed.com

Beach Cams:
www.evsmartin.com (Martin County)
www.windjammerresort.com/webcam.html (Lauderdale-by-the-Sea)

INDEX

Page numbers in italics indicate sidebars, illustrations, and photographs.